Sydney Lupton

Notes on Observations

Sydney Lupton

Notes on Observations

ISBN/EAN: 9783337337650

Printed in Europe, USA, Canada, Australia, Japan

Cover: Foto ©berggeist007 / pixelio.de

More available books at **www.hansebooks.com**

NOTES ON OBSERVATIONS

BEING

AN OUTLINE OF THE METHODS USED FOR DETERMINING THE MEANING AND VALUE OF QUANTITATIVE OBSERVATIONS AND EXPERIMENTS IN PHYSICS AND CHEMISTRY, AND FOR REDUCING THE RESULTS OBTAINED

BY

SYDNEY LUPTON M.A.

London
MACMILLAN AND CO., Limited
NEW YORK: THE MACMILLAN COMPANY
1898

All rights reserved

PREFACE.

THE following pages are intended to assist a beginner in realizing the value of the quantitative results, which he himself and others have obtained, in physical and chemical experiments.

Questions involving higher mathematics and disputed proofs are omitted as unsuited to the object of the book; but it is hoped that the references given may be of assistance to those who desire to pursue the subject more thoroughly.

No attempt is made to give a complete account of the method of least squares, but only an outline sufficient to enable one unacquainted with mathematics to use it for practical purposes.

S. L.

CONTENTS.

CHAPTER I.
IDEAS, - - - - - - - - 1   PAGE

CHAPTER II.
REASONING, - - - - - - - 7

CHAPTER III.
FALLACIES, - - - - - - - 13

CHAPTER IV.
LAWS OF NATURE, - - - - - 18

CHAPTER V.
CAUSE AND EFFECT, - - - - - 22

CHAPTER VI.
OBSERVATION AND EXPERIMENT, - - - 31

CHAPTER VII.
Units and Dimensions, - - - - 36

CHAPTER VIII.
Averages, - - - - - - 41

CHAPTER IX.
Differences, - - - - - - 46

CHAPTER X.
Interpolation, - - - - - 51

CHAPTER XI.
Mensuration, - - - - - - 56

CHAPTER XII.
The Use of Tables, - - - - - 60

CHAPTER XIII.
Errors, - - - - - - 66

CHAPTER XIV.
Means, - - - - - - 71

CHAPTER XV.
The Law of the Frequency of Errors, - 74

CONTENTS. ix

CHAPTER XVI.
The Weight of Observations, - - - 84

CHAPTER XVII.
The Method of Least Squares, - - - 88

CHAPTER XVIII.
Conditioned Equations, - - - - 90

CHAPTER XIX.
General Formulae, - - - - - 92

CHAPTER XX.
The Deductive Method, - - - - 95

CHAPTER XXI.
The Expression of Results by Graphical Methods, - - - - - - 97

CHAPTER XXII.
The Expression of Results by Empirical Formulae, - - - - - 104

CHAPTER I.

IDEAS.

WHEN a man in his ordinary and normal condition considers how he thinks, he is conscious of two different entities. He feels that he himself possesses the power of thinking,[1] or that he has an individuality distinct from everything else. This self in philosophical language constitutes the ego, the I-myself, the subject. He feels also that he is affected by things outside himself, which constitute the non-ego, the not-I, the object, the environment.

It is impossible to state accurately in what individuality consists, or to give any satisfactory definition of it. Though it is much influenced by the body, it does not entirely depend upon it, since the body may be considerably changed by time, accident, or disease without loss of individuality. If individuality be defined as the possession of the power of thinking and of the consciousness of exercising that power, the ordinary meaning of the term is much restricted, and individuality is denied to infants, to persons asleep, unconscious, or cataleptic, and to the mentally alienated. In some rare but extremely interesting cases of dual or triple consciousness[2] the same bodily form presents at different times two or three distinct mental personalities, each of which has no apparent connection with the other, except that it influences the same body.

It may perhaps be assumed that every being which is conscious of the power of thinking has a separate individuality; but it must always be borne in mind that the converse—every being which is not endowed with the power of conscious cerebration is not an individual—is probably untrue.

So far as is known all thinking requires the action of some portion of the brain, and if certain portions of the brain are injured or removed, corresponding powers of thought are lost. That which thinks, whether it be identical with the individual or not, is spoken of as the mind, and is supposed to be localized in the brain.

Any presentation to or impression upon the mind of a subject is called an idea. An idea is said to be concrete when it is supposed to have a co-relative in the external world, and abstract when it is supposed to have no such co-relative. We believe that a class of animals represented by the word dog exists, and therefore dog is a concrete idea; a mathematical straight line does not exist, and therefore the idea of a straight line is an abstract one.

Ideas may possibly be produced by the action

of the mind itself, by the action of the external world through the bodily senses, that is, by sensations,[3] or by remembrance of, and reflection upon, those sensations. It must be carefully noted that external objects may act upon the body of the subject without affecting the mind. Such actions, of which automatic and reflex actions are familiar instances, are not sensations as above defined.

Everything exterior to ourselves can only be perceived and appreciated by our own senses, we know the environment only by its effect upon the I. Hence we have no means of determining how far our sensations really represent the phenomena which are assumed to produce them, or how far the same phenomena produce similar sensations in different individuals. When we say that we reason from facts, we really mean that we argue from ideas which more or less truly represent phenomena.

The familiar phenomenon that the earth revolves on its axis, owing to an incorrect interpretation of the sensations, produced for many centuries the false idea that the sun goes round the earth.

It has long been a vexed question, if any, and if so what, ideas are innate,[4] or due to the mind itself. Most of the ideas which are by some assumed to be innate are abstract, and are of more importance in ethics and theology than in natural science. Many of the older philosophers held that all ideas are derived directly or indirectly from experience; but the question has recently passed into a fresh stage, owing to the general acceptance of the doctrine of heredity.[5]

The theory that like produces like is now very generally admitted so far as bodily form and function are concerned, and very striking applications of it are shown in the improved breeds of our domestic animals. There seems no reason for excluding the brain, or those portions of it which are the organs of thought, from the operation of the general law,. hence we should expect to find somewhat similar mental characteristics transmitted from parents to offspring. In many cases facts seem to bear out the theory, though the complication of the circumstances often renders the interpretation doubtful.

Many cases of inherited instincts among the lower animals are well known. Ducklings hatched by a hen take readily to the water. A young pointer often points the first time it has seen game. A puppy, deprived of its begging mother, has been known to beg. It is much more difficult to obtain trustworthy evidence in the case of man,[6] but such varied tendencies as drunkenness, powers of arithmetical calculation, and facility in learning languages seem to descend. It may, of course, be argued that these are merely tendencies, and that no definite idea can be proved to be transmitted in the case of man.

Speaking generally, any irritation of a sensorial nerve is conveyed through the nerve substance to the brain, where it produces a sensation which evokes an idea.

IDEAS.   5

Thus the radiant energy from a burning candle passes as a series of vibrations through the ether until it falls upon the pupil of the eye, producing an image of the flame upon the retina. The irritation of the retina is transmitted through the optic nerve to the brain, and produces the sensation of light. The idea of a lighted candle is produced by many sensations aided by memory and reflection. If the light be very bright the sensation is that of pain, and the eyelid closes automatically without any act of will on the part of the subject.

Our sensory apparatus is at best very imperfect, thus our sensations are liable to deceive us. We also frequently err in interpreting a sensation into an idea. A sharp blow on the eyeball causes a sensation which is interpreted into the idea of a flash of light, until the false idea is corrected by a subsequent sensation of pain.

In making any observation, therefore, it is necessary to be on our guard against self-deception in our sensations and in the incorrect representation of them as ideas.

REFERENCES AND NOTES.

1. Descartes (1596–1650): "Cogito, ergo sum."
2. Binet: *Alterations of Personality*. London, 1896.
3. Sensation is frequently used for what is here called the idea produced by it. For details, see Huxley: *Elementary Physiology*. Bernstein: *The Five Senses of Man*.

4. Plato, Aristotle, Descartes, Leibnitz, Kant held that some ideas are innate; Bacon, Locke, Hume, James Mill, J. S. Mill that all ideas are derived directly or indirectly from experience.

5. C. Darwin (1809-82): *The Origin of Species* (1859). Inheritance is discussed in Chapter i., and Instinct in viii. See also H. Spencer: *Biology*, viii.

6. F. Galton: *Hereditary Genius*, 1869.

CHAPTER II.

REASONING.

WHEN ideas are coordinated so that similarities and differences between them are perceived, the statement of the relationship is called a premiss.[1]

Premisses may be derived from three sources:

(i.) Subjectively, from our own minds. Some ideas may be innate, hereditary, or derived unconsciously from the environment. Some of these premisses, such as that of our own personality, are among the most certain which we know; others are extremely doubtful, since they depend upon the evidence of one mind only, our own.

Certain definitions and axioms, which may be due to ourselves or others, are so thoroughly accepted by our own minds that they seem to be inherent in them. They are almost all abstract, generally mathematical, and universally admitted by all capable of understanding them. Thus, "if equals be added to equals the wholes are equal," can hardly be disputed.

(ii.) Objectively, by the observation of external objects. Premisses of this class vary much in value with the special facilities, natural aptitude, and adequate training of the observer. A keen-eyed man can see two of the moons of Jupiter, extraordinary sight is required to see four, a first-rate and well-situated telescope is needed to obtain any knowledge of the fifth.

(iii.) Authoritatively, by communication from an intelligence supposed to be better informed than our own.

The intelligence may be human, and, as we suppose, more or less similar to our own; or superhuman and quite different from ourselves. Since by common consent no natural knowledge has been directly revealed, natural science, as such, has to deal only with human intelligences.

The proportion of our knowledge gained by this method tends to become larger the better instructed we are, since our individual observations and reflections must be very few in comparison with the total number accumulated by mankind.

It must be always carefully borne in mind that the authoritative intelligence, if human, has itself gained its information in one of these three ways. Hence, other things being equal, the validity of a premiss obtained by this method is less than that of one obtained by either of the other methods, since there is additional chance of misunderstanding, and error in the transmission of the information from the intelligence to ourselves. On the other hand this chance may be more than counterbalanced by the superior knowledge, skill, or appliances of our informant. Any ordinary person would assume a value given in the *Nautical Almanac* to be more accurate than one obtained by his own observation.

Generally speaking, authority, especially if it be ancient, is accepted with too little question. The race is older than it was in the time of our ancestors, hence in many respects we are wiser than they were. A schoolboy can now use knowledge and appliances which were quite beyond the reach of Newton. Further, the conception that accuracy in quotation and description is an essential duty is of late growth. It is very doubtful if the speeches in Thucydides were ever delivered, and tolerably certain that the crimes of the Caesars were heightened by the art of Tacitus.

An authority may give the reasons on which his statement is based, when we are generally able, if sufficiently instructed, at least to see that there are no glaring faults in the argument. Far more commonly, however, we have to deal with a statement for which no reasons and no further evidence is given. In such a case we can only form a more or less probable guess as to the truth of the statement based upon its inherent probability or improbability, the good faith of our informant, and the means at his disposal for ascertaining the truth.

The statement that the twenty-second figure in the value of $\cos \frac{2\pi}{7}$ is 0 will be readily believed.

The statement that by the methods of Euclid a construction had been found for squaring the circle, trisecting the angle, or duplicating the cube would be unhesitatingly disbelieved.

Fermat claimed to have discovered a general proof that no integral solution of $x^n + y^n = z^n$, if $n > 2$ and an integer, can be found. During 330 years, the statement has been shown to be true for some half-dozen values of n, but the general proof is still unknown. How far is Fermat's statement credible?

When the premisses have been obtained by either of these methods, they may be reasoned upon by induction or by deduction. In inductive reasoning a number of individual instances are compared, and a general statement[2] or proposition is made which includes them all, and also other similar instances which have not been examined. Induction then is the operation of discovering and proving general propositions.[3]

When a proposition is found to be true of every member of a class and then applied to the whole class by a kind of short-hand, some logicians consider that a perfect induction has been made; others, that there has been no induction at all, but only a colligation of instances. The cases in which we can examine every possible instance are comparatively few, and may for the purposes of natural science be neglected.

In general we can only select a very limited from a very large number of instances, and to a certain extent assure ourselves that they are fairly representative of all possible cases of the same kind. From these special instances we frame a general proposition which includes them all, and is assumed also to include all instances of a like kind, past, present, and future. We then verify the general proposition by finding fresh instances in which it applies, and by making deductions from it, and proving that the results of the deductions agree with facts.[4]

In deductive reasoning, particular results are shown to follow from definitions or from general statements.

Having examined gold, silver, copper, iron, etc., we may, by induction, lay down the general statement, that at ordinary temperatures all metals conduct electricity well. A new metal, *e.g.* gallium, is discovered. We argue deductively that gallium, being a metal, will be found to conduct electricity well. We make the experiment, and find that our deduction agrees with the fact. So far as this one instance goes, we have verified our original induction, and our belief in its validity is so far justly increased.

No conclusion can be more certain than the premisses from which it is derived. Since, except in mathematics, in which the definitions are precise and generally admitted, in almost every case we must be doubtful as to the exact correspondence of our ideas with the actual facts, very few of our conclusions are absolutely certain. In general we have to accept and act upon conclusions which are only more or less probable. Hence, as Butler says,[5] "Probability is the very guide of life."

A conclusion which we recognize as not certain, but only probable, is expressed by the word belief. "Belief,[6] assent, or opinion, is the admitting or receiving any proposition for true upon arguments or proofs that are found to persuade us to receive it as true, without certain knowledge that it is so."

REFERENCES AND NOTES.

1. Books on Logic are very numerous, but the older authors pay little attention to induction, or to examples in science.

S. Jevons: *Logic* (Science Primer), a clear outline.
J. S. Mill: *A System of Logic.*
S. Jevons: *The Principles of Science.*
J. Venn: *Inductive Logic.*

2. Statements may be expressed at full length in words, but especially if they are quantitative, conciseness of expression and facility of manipulation are frequently gained by the use of symbols. In addition to the use of symbols as a general expression for quantities, they are also used to express operations upon quantities. This conception has led in mathematics to the valuable "calculus of operations," in logic to the "symbolic logic" of Boole, and in chemistry to the "ideal chemistry" of Brodie.

3. Mill: *Logic*, iii. 1.

4. Laplace: *Essai Philosophique.* Des divers moyens d'approcher de la certitude.

5. *Introduction to the Analogy of Religion*, p. 3.

6. Locke: *An Essay concerning Human Understanding*, iv. 15, 3.

CHAPTER III.

FALLACIES.

BESIDES uncertainty due to the constitution of our own minds, our conclusions, whether they be derived from our own observations and reflections or from the authority of others, are liable to errors, which may lurk unperceived in the premisses, in the reasoning, or in the conclusion. These errors are termed fallacies, and occupy a prominent position in every work on logic.

Some fallacies are more common in inductive, and others in deductive reasoning. The latter were discussed with great care and subtilty by writers of the Middle Ages; but the discussion followed certain classical lines, and was hampered by medieval terms which are now unfamiliar.

The most common cause of error in inductive reasoning is neglecting to consider sufficiently numerous and sufficiently widely distributed instances. We must assure ourselves, as far as possible, that the instances selected cover the *whole* class of cases with which we are dealing.

If after examining the instances of sound, heat, light, and electricity, we lay down the general proposition—all forms of energy take a finite time to traverse space, the induction is imperfect, and, so far as we at present know, untrue—the case of gravitation has been neglected.

Again the general proposition may be expressed inaccurately, because the true points of similarity between the instances have not been adequately grasped. These are called by Mill fallacies of generalization.

Until early in 1895 the general statement that all elements are capable of entering into chemical combination seemed a nearly certain induction. The discovery of argon and helium proved the fallacy of it.[1]

In deductive reasoning "The great source of fallacies is confusion, the great safeguard against them is to think and express oneself clearly."[2]

The premisses may be unduly assumed, or not clearly expressed. A premiss may assume or depend upon the conclusion. The premisses may have only an apparent or verbal, and not a real or necessary connection.

In the course of the reasoning, especially if it be long and not fully expressed, an untrue premiss may be tacitly assumed, a word or phrase may be used in different senses or in one place with reservations, and in another without.

The conclusion may not follow from the premisses

because the ground of the argument has been consciously or unconsciously shifted, which often occurs in stating an opponent's case, or because an irrelevant conclusion, such as a jest, a personal attack, or an appeal to sentiment, has been introduced by accident or on purpose instead of the true conclusion.

It is often assumed that failure to prove a conclusion proves the opposite. This is of course never the case, though if a well-informed and skilful reasoner fails, there is some ground for supposing that the proof, if it be possible, is not easy. This illegitimate assumption must not be confused with the legitimate method so often used by Euclid. A premiss is assumed and shown by correct reasoning to lead to a false conclusion *quod est absurdum*, hence it follows that the assumed premiss is itself untrue. Care must of course be taken that the error does not lie in the reasoning.

Perhaps the most widely spread and insidious fallacy is from doubtful premisses to reach a conclusion, which is assumed to be absolutely certain, instead of a more or less probable belief.

Every additional doubtful premiss in a chain of evidence lessens the probability of the conclusion, so that comparatively few doubtful premisses render the result of little validity. If there be only four independent premisses, the probability of each of which is $\frac{3}{4}$, the validity of the conclusion, so far as it depends upon this chain of reasoning, is only about $\frac{1}{2}$, or it is an equal chance that it is true or false.[3]

Since all beliefs are merely more or less probable, they should be held subject to revision on fresh evidence or further consideration. The majority of mankind, however, cherish most closely those beliefs which they hold on the most doubtful evidence, and from their own point of view wisely refuse to listen to any arguments against them.[4]

The remarks made so far refer to reasoning in general. The use of reasoning for the purposes of physics and chemistry is but little obnoxious to some forms of errors which have been mentioned, while it is very liable to some forms, *e.g.* arithmetical,[5] which are more or less peculiar to itself.

REFERENCES AND NOTES.

1. Early in 1895 at least 73 elements were known, all of which would enter into combination. The chance that the next element discovered would enter into combination was by Bayes' theorem (cf. Bertrand, *Probabilités*, p. 172) $\frac{73+1}{73+2}$, or only $\frac{1}{75}$ against argon entering into combination. This instance shows with what doubt and care such calculations must be applied in practice.

2. J. S. Mill: *An Examination of Sir W. Hamilton's Philosophy*, p. 525.

Whately: *Logic*. The chapter on fallacies is amusing and interesting.

Shute: *A Discourse on Truth*. More technical.

3. If there be n independent premisses, the chance in favour of each of which is p and against it q, and if

$$\frac{p_1}{p_1+q_1} \times \frac{p_2}{p_2+q_2} \times \ldots \times \frac{p_n}{p_n+q_n} = \tfrac{1}{2},$$

the chances in favour and against the conclusion are equal, or such a chain of reasoning, even if technically correct, adds no weight to whatever other arguments there may be in favour of the conclusion.

4. In this connection Lecky, *The History of Rationalism*, will be found most interesting. It is difficult for one brought up in a more enlightened atmosphere to realize the terrible struggle portrayed in J. A. Froude's *The Nemesis of Faith*.

5. Besides very frequent and unexpected errors, arithmetical results frequently pretend to an impossible degree of accuracy, cf. S. Lupton: "The Art of Computation for the Purposes of Science," *Nature*, Jan. 5th and 12th, 1888.

CHAPTER IV.

THE LAWS OF NATURE.

THE word law is used with two entirely different significations, and is therefore frequently a source of misconception and fallacy.

In common usage, a law means an edict or rule imposed by superior power, which either must be obeyed, or if disobeyed, deserves, and possibly entails, pains and penalties.

In science, however, the meaning of the word is quite different. "Laws of Nature"[1] are simply propositions stating uniformities which have been observed in the relations of phenomena, and may therefore be expected to obtain in similar cases.

Up to the end of last century it was always observed that a substance which had a metallic lustre was several times heavier than an equal volume of water. The discovery of potassium, with a high metallic lustre, but with a density much less than that of water, broke no authoritative edict, but merely showed that the old law was not sufficiently wide.

THE LAWS OF NATURE.

The study of natural science consists in the observation of, and reflection upon, differences and changes in two manifestations which are spoken of as matter and energy.

"We are acquainted with matter only as that which may have energy communicated to it from other matter, which may in its turn communicate energy to other matter. Energy, on the other hand, we know only as that which in all natural phenomena is continually passing from one portion of matter to another. Energy cannot exist except in connection with matter."[2]

Matter and energy, though they may change their form in a great variety of ways, so far as our experience extends, can never be created or destroyed.[3] Hence every change is due to some form of previously existing energy, or every effect is due to a cause. When two changes always take place together, it may in general be assumed that the first is the cause of the second, or that both are due to the same cause.

There are only two general methods by which a scientific law or any other general proposition can be established, of which the second is far less definite and far more liable to error, but far more generally applicable than the first.

In mathematics, and in some of the better explored parts of the experimental sciences, we may lay down certain definitions or theories, and by deductive reasoning arrive at further results. Of such reasoning Euclid is a typical instance. Maxwell, assuming the laws of motion, gravitation, and the viscosity of gases, showed that a ring of gas could not continue to rotate round a planet. The ring of Saturn has recently been proved by observation to consist of solid particles.

In the great majority of cases, owing to our ignorance of general laws, we are obliged to use inductive reasoning. By passive observation of phenomena presented to us, or by active observation under conditions brought about by ourselves, that is, by experiment, we collect a number of instances of a uniformity. Reasoning upon these instances, we make an hypothesis which includes them all. If this hypothesis agrees with all instances already known, and especially if it agrees with fresh instances as they arise, we consider that it is more or less probably true, and call it a theory. The hypothesis must be discarded and another formed if any instance is found to contradict it. When a theory has stood the test of time, and if deductions made from it are found to agree with observation, it is considered to be very probably true, and spoken of as a law of nature.

The motions of the planets are accounted for if an original impulse was given to them, and if every other particle of matter attracts every other particle directly as the product of the masses, and inversely as the square of the distance between them. The theory of universal gravitation was found to account for the tides and for the shape of the earth; Adams and Leverrier deduced from it the existence of the planet now known as Neptune. Few truths are held to be more certain than the Newtonian law.

REFERENCES AND NOTES.

1. H. Spencer: *Essays*, iii. 81. "Of Laws in General."
2. J. Clerk Maxwell: *Matter and Motion*, p. 93.

The conception of matter and energy as two separate entities is probably the most simple for the beginner. He will find later that all the different properties assigned to various kinds of matter may be due to energy, which produces different modes of motion in one medium— the ether. Going a step further back, energy and ether seem so necessarily connected in thought, that it is impossible to distinguish between them. Hence all the forms of matter, and all the forms of energy, may be simply presentations of various modes of motion of the ether.

3. H. von Helmholtz: *Popular Lectures*, p. 277. "On the Conservation of Force."

CHAPTER V.

CAUSE AND EFFECT.

THE connection between causes and the effects which follow them has been discussed by philosophers from time immemorial, but no general agreement has yet been reached.

Metaphysicians attacked the question from the side of the human will, which they assumed to be the cause of various effects. Thus volition can produce an idea in the mind and motion in the body. From the human will they have argued by analogy that a Divine Will exists, which has planned out the universe, and is the ultimate cause of all phenomena. The analogy is, however, so imperfect that, valuable as it may be as an illustration, the use of it as an argument seems absurd.

To consider only two points of difference. It is very doubtful if human volition can evoke any idea, the separate portions of which are not supplied by memory. Though I can call up the idea of a unicorn, the separate concepts of which the idea is composed are due to my own observation, or to communication from others. The Divine Will is presumed to be self-originating.

The connection between volition and muscular action is still very obscure. Each act of will appears to be produced by, or to produce a change in, a special portion of the brain, and this stimulus is conveyed by one or more nerves to the special muscles concerned. Muscular tissue is apparently of at least two different kinds; the one variety which, with the exception of the heart, is generally unstriped, acts more or less automatically, though it may be stimulated by external means. Voluntary muscles are generally striped, and, when acted upon by a stimulus conveyed through a nerve from the brain, or from an external source, contract or expand, producing motion in the corresponding parts.

The contraction or expansion of a muscle requires a certain amount of energy which is supplied by the oxidation of the blood in vessels lying in the interstices of the muscular fibres. The oxygen is supplied to the blood in the lungs, and the oxidizable constituents by the digested food. The products of the combustion are ejected.

Each muscle then may be regarded as an engine which converts the potential energy of the food and oxygen into kinetic energy. The amount of energy which must be supplied in any given case is, so far as we can tell, exactly the same whether an act of will has been performed or not. Food is required to keep up the temperature of the body, to repair the waste of tissue, and to perform external work. The human will only produces mechanical effects after a long series of changes, each of which is limited by extremely stringent conditions. Hence we are not justified in assuming that the act of will is *the* cause of mechanical action and not merely a condition, possibly not even a necessary condition, since a very minute electric stimulus may produce a very considerable muscular action. The Divine Will is assumed to act directly without conditions, and to Itself supply the energy required to produce the effect. It is thus conceived to be totally distinct, not merely in quantity but also in quality, from the human will, and no analogy can be pressed.

The question of cause and effect has been further complicated by the introduction, in many instances, of interminable discussions as to the limits of free-will and necessity. Human intelligence is unable to reconcile a perfect plan, omniscience, and omnipotence on the part of the Divine Will with freedom of choice and action, and therefore with moral responsibility on the part of the human will.

So far as physics and chemistry are concerned all questions of the action of will may be omitted, and the inquiry restricted to inanimate causes, and thus rendered more comprehensible.

The language of some philosophers has been affected by the loose popular use of terms. The word "cause" has often been applied by analogy in cases to which it is not really applicable. It has also been used to express the absence of a condition which would have prevented the occurrence. Strictly speaking, non-existence cannot be the cause of anything. Hence any accurate definition of the relationship of cause and effect must much restrict the popular use of the words.

Before any effect can be produced several or many conditions must be present, the most marked of these conditions is picked out and spoken of as *the* cause of the effect, though in reality several of the little noticed conditions may be as necessary for the production of the effect as the one selected as the cause.

So far as we know every change stands in the relation of effect to some change which has preceded it, and in that of cause to some change which follows it. Hence every phenomenon may be compared to a link in a chain which stretches backwards and forwards beyond our ken.

When we attempt by the aid of experiment and reflection to pass back from the phenomena we observe to the causes which produce them, we are able gradually to include several less general uniformities under one more general law, and to ascribe a greater number of phenomena to one cause. However far we may be able to carry this process, we are stopped at last by an impassable boundary which recedes slowly before the greater knowledge of the race. On the other hand, when we attempt to trace the effects produced by any cause, we soon find ourselves in a tangled skein, owing to effects due to causes external to the one we are trying to investigate. In any attempt to discuss ultimate or final causes natural science ceases to be experimental, and passes into metaphysics.

Such different phenomena as the shape of the earth, the falling of a stone to it, the motion of the tides on its surface, the motion of the moon round it, the fact that it does not fly to pieces, and that it is mountainous, may be said to be caused by the law of universal gravitation, though the phrase is apt to mislead unless the meaning of "law" be remembered. But we are quite unable to discover the cause of gravitation, why it obeys the law of the inverse square of the distance, and why it differs from all other known forms of energy in its infinite velocity of propagation. The highest mathematics are unable to solve exactly the motion of three nearly equal heavy particles. We are unable to say when and where the next chain of mountains will be formed. We can only approximate to the true theory of the tides, and the true motion of the moon.

Some philosophers regard universal sequence as the true and only test of causation. In the words of Mill,[1] "To certain facts, certain facts do, and, as we believe, will continue to succeed. The invariable antecedent is termed the cause, the invariable consequent the effect."

Stanley Jevons[2] objects to the rigidity of this definition : "If a cause is an invariable and necessary condition of an event, we can never know certainly whether the cause exists or not. To us then a cause is not to be distinguished from the group of positive or negative conditions, which with more or less probability precede an event."

Both definitions are so wide that some reservations must be understood. The stars disappear before the sun rises; is their disappearance the cause of his rising? Can we correctly speak of the cause of an event which has happened only once, and is the title, "On the causes of the present discontents," a misnomer? The words "positive and negative" require definition; it seems unnecessary to include among the causes of an event all the conditions which, if present, might with more or less probability have prevented the occurrence. Such a tedious enumeration would too often divert attention from the true points at issue.

A second condition, frequently assumed to be necessary, is a material connection between the body assumed to be the cause and the body affected, "since nothing can act where it is not." Nothwithstanding the high authority of Newton,[3] this condition is by no means universally accepted. So far as experiment at present goes, gravitation acts immediately and directly across space with infinite velocity.[4] It is difficult to conceive the constitution of an elastic ether, which is truly continuous with no space between the particles.[5]

A third condition is now universally admitted, and since it includes the first and possibly the second condition, it may be provisionally used as a definition of the relation of cause and effect.

When more energy passes from one material system to another than returns from the second to the first and produces an evident change in the second system, the first system is said to be the cause of the effect in the second system, This definition suffices for the purposes of physics and chemistry, but restricts the use of the terms within far narrower limits than is usual.

A few remarks and examples may render some points in this very difficult subject more clear. A comparatively small cause by setting free potential or stored up energy may produce very great effects. A single spark may cause the explosion of a barrel of gunpowder; the amount of energy set free in such a case is practically infinite in comparison with that transferred from the cause.

It is now universally admitted that every effect must be produced by a cause, but in many cases the causes are imperceptible or doubtful. Substances pass into isomorphous forms with the evolution or absorption of heat apparently by mere lapse of time, *e.g.* viscous or prismatic into octohedral sulphur, yellow into red mercuric iodide. The action of traces of impurities such as water vapour in assisting chemical combination is little understood. Traces of impurity also greatly decrease the conductivity of metals, and alter the properties of steel. More boys than girls are born every year in the proportion of about 104 : 100, and all the ingenuity of Laplace failed to discover a cause. Generally speaking it is more difficult to determine causes from their effects, than effects from their causes.[6]

When two bodies are separated by media transparent to a form of energy possessed by either or both of them, that form of energy passes from one or each of them to the other in equal or unequal amounts. Every substance known is transparent to gravitation, and all known bodies show that form of energy, two bodies always attract one another equally.

If two bodies are separated by air, radiant heat perpetually passes from each to the other. If the temperature of the two bodies be the same, the same amount of energy passes in each direction, no evident change takes place, and we do not consider either body as the cause of the temperature of the other. If, however, the one body be a red-hot cannon ball, and the other a lump of ice, more energy passes from the ball to the ice than in the reverse direction. Hence we are justified in saying that the ball is the cause of the melting of the ice, but we are not strictly correct in saying that the ice is the cause of the cooling of the ball.

In the absence of the atmosphere, a raindrop falling from the height of a mile would strike the earth with a velocity of 580 feet per second, and a shower would be as destructive as a torrent of shot. Strictly speaking, the air is a condition and not a cause of the harmlessness of showers. No energy passes from the air to the drops; but the energy, as it is produced by the action of gravitation, is more or less expended in tearing the air asunder, and in heating the air and drops.

REFERENCES AND NOTES.

1. Mill: *Logic*, III. v.
Thos. Brown: *Lectures*, vii.; also *Essay on the Relation between Cause and Effect.*
Whewell: *History of Scientific Ideas*, Book III., ch. ii.
2. Stanley Jevons: *Principles of Science*, II. xi. 220.
3. *Correspondence of Richard Bentley*, i. 70.
Lord Kelvin: *Popular Lectures*, ii. 538.
4. Herschel: *Familiar Lectures*. "The Sun," 12, and note.
5. O. Lodge: *Modern Views of Electricity*, 358.
6. Every work on probability gives an account of the attempts which have been made to find an expression for the relative probability of different causes from the given probabilities of their effects.
Bertrand: *Calcul des Probabilités*, vii.
The use of the inverse form of the theorem of James Bernoulli, and of the theorem of Bayes is open to very grave doubts, and seems to be of little practical utility since the conditions required for their valid application are rarely if ever fulfilled. Cf. Boole: *Laws of Thought*, 365.

CHAPTER VI.

OBSERVATION AND EXPERIMENT.

WHEN we notice a change in any phenomenon, or compare any quality of two objects, we are said to make an observation. At first observations were purely qualitative, and merely denoted the presence or absence of certain phenomena, or rough impressions of shape, colour, and other qualities. Gradually, however, observations became more precise, and were extended to size and number, or became quantitative.[1] At first numerical estimates were made directly, as when a man is said to weigh twice as much as a boy; in modern times, measurements are made by the aid of instruments with increasing accuracy, and the quality or change observed is expressed in terms of a unit by a numeric;[2] the man is said to weigh twelve stone, and the boy six.

The accuracy attained in any case depends upon the special circumstances of the observation, the care with which the operations are conducted, and the personal skill of the observer. There is a natural tendency to overestimate the accuracy and value of observations upon which one has expended much time and trouble, and this common failing must be carefully guarded against. The exact numeric which connects any observed quantity with the unit can never be obtained except by accident, and when it has been obtained, there is no means of ascertaining the fact. The difficulty of observing, and the number of necessary corrections increase very rapidly, as the accuracy aimed at is greater; in many delicate measurements corrections have to be applied which are larger than the quantity to be measured. Hence very few of the numerics found by observation or experiment are reliable beyond the sixth significant figure, and the great majority to a far smaller extent.

It is well to bear in mind the ironical answer of Dulong, when asked why he gave indices of refraction to eight figures, when his results only agreed to three: "I do not see why I should suppress the last decimals, for if the first are untrue, possibly the last are correct."

It is useless, wearisome, and misleading to extend the numeric beyond the first doubtful figure, all the rest should be suppressed or replaced by ciphers.

If the phenomena which we wish to investigate are or can be reproduced near to us, we are generally able to alter the conditions under which they occur, and thus render our observations more general, more definite, or more precise. We are then said to make an experiment.

OBSERVATION AND EXPERIMENT.

After observing a rainbow, we may experiment on bows formed by spheres of various transparent substances in the laboratory, and show that the position of the bow depends upon the refractive index of the substance by which it is formed, and hence that our theory of its formation is probably correct, and applies to all transparent substances, as well as to water.

We may observe the effect of different soils upon the growth of wheat, and make our observation more definite by the addition of one substance such as guano to a portion of a plot. If the other circumstances remain the same, any increase in yield must be due to the guano.

The velocity of a freely falling body soon becomes so great, that it is difficult to determine the space passed through in a given time with any accuracy. By attaching a heavy body to a grooved wheel which was caused to run down an inclined wire, Galileo obtained much better results.[3]

When great accuracy is desired, each observation must be repeated a considerable number of times, if possible, by several different persons, since all observers are liable to personal errors. Some read late or early, some read high or low, some are more or less colour blind. In arranging experiments it is most important, so far as possible, to measure only one property or one change at once, and to avoid great size or complexity in the apparatus. It is well also to use different samples of the substance under examination obtained from different sources, and at least two different methods and sets of apparatus. We may thus avoid or allow for errors due to the personal defects of the observer, to the impurity of the substance, and to imperfection in the apparatus.

Many persons had weighed samples of nitrogen obtained from the atmosphere before Lord Rayleigh, by weighing samples obtained from different sources, noticed a difference in weight which led to the discovery of argon.

Dumas based most of his determinations of atomic weights on weighed quantities of pure silver. Hence a certain doubt attaches to almost all his results, since he did not get rid of, or allow for the oxygen occluded by the silver. Stas in most cases used several different methods to determine each atomic weight.

However carefully observations may be made, they cannot be accurate or comparable with those of other observers, unless the units in terms of which the results are expressed are accurate and comparable with those used by others. Hence before any quantitative observations are made, the units in terms of which they are to be recorded must be carefully considered.

REFERENCES AND NOTES.

1. Tylor: *Anthropology*, p. 309.
2. A numeric is any abstract numerical expression which may be whole, fractional, or mixed. Thomson: *Arithmetic*, p. 4.
3. Galileo (1564-1642), the founder of modern mechanics, made many discoveries in astronomy.

If the mass of the particle and wheel be m, the angle between the wire and horizon θ, and the acceleration due to gravity g, resolving perpendicular to the wire $mg\cos\theta$ produces pressure on the wire but no motion, and along the wire $mg\sin\theta$ is an accelerating force causing motion in the system. But by making θ, and therefore $\sin\theta$, small, we may make the accelerating force as small as we please, and therefore the motion as slow as we please.

4. Compare the large apparatus for determining the constant of gravitation G, used by Cavendish and Baily, with the far smaller and more accurate one used by Boys.

CHAPTER VII.

UNITS AND DIMENSIONS.

THAT measurements may be exact the unit used must not vary with change of time or place, and must be easily reproduced or represented by accurate and permanent copies.

Some units[1] are said to be natural, because they are, as it were, presented to us ready-made; when available they are generally the most convenient. In many cases no natural unit exists, and we are obliged to construct and define an artificial one.

Astronomers use as their unit of time the sidereal day, or the interval between two successive passages of a star across the meridian, and divide it into hours, minutes, and seconds. By noting the times of two transits of the same star an astronomer can readily determine the error of his clock. The ordinary unit of time, the mean solar day, is not a natural unit, since it is obtained by adding nearly four minutes to the sidereal day; it is, however, easily

UNITS AND DIMENSIONS.

and readily obtained at all places in telegraphic communication with Greenwich Observatory.

Though no available natural units of length or mass exist, artificial units of these quantities have been constructed and legally recognized in all civilized countries. The best material has been found to be an alloy of platinum with one ninth of its mass of iridium. The standard metre is an X-shaped bar of this material with fine lines traced on the depressed part, the distance between two of these lines, when the bar is at the temperature of melting ice, constitutes the metre. The standard kilogram is a cylinder of iridio-platinum, the diameter of which is equal to the height. Both these standards are preserved at Breteuil, and copies of them have been sent to all civilized governments. England and the United States still use the yard and pound as units.

Though some other fundamental units, such as those of angle, temperature, and illumination, are required, it is a great convenience as regards both cheapness and simplicity of calculation, to use as few fundamental units as possible, and to measure all quantities in terms of units derived from them.

Instead of constructing new fundamental units to measure area and volume we may use for the one a square surface, a metre along each edge, and speak of a surface containing so many square metres; and for the other a cube, a metre along each edge, and speak of so many cubic metres or steres.

The frequent use of derived units has brought into prominence the modern theory of dimensions.[2]

If any quantity Q be measured in terms of a length L, a mass M, and a time T, so that

$$Q = L^a M^b T^c,$$

the quantity Q is said to be of the dimensions a in length, b in mass, c in time. Thus the stere is of the dimension 3 in length, 0 in mass and time.

The following are the dimensions of the chief limits used in mechanics:

Length,	L^1	Time,	T^1
Area,	L^2	Velocity,	L^1T^{-1}
Volume,	L^3	Acceleration,	L^1T^{-2}
Mass,	M^1	Work, Energy,	$L^2M^1T^{-2}$
Density,	$L^{-3}M^1$		
Momentum,	$L^1M^1T^{-1}$	Power,	$L^2M^1T^{-3}$
Force,	$L^1M^1T^{-2}$		

The theory of dimensions renders great assistance in passing from one set of units to another. Thus, if the unit of length be kept the same, and the unit of time be changed from seconds to minutes, the unit of acceleration is decreased, not 60, but 60×60 times; and hence a quantity measured in terms of the first set of units must be multiplied by 3600 to refer it to the second set of units.

For scientific purposes it is perhaps most convenient to use only one unit of each kind, and to express any quantity in terms of it, and of a numeric composed of not more than one digit of whole numbers and a decimal multiplied by a positive or negative power of ten. Thus the velocity of light, according to Cornu, is 3.004×10^{10} cm. per sec., and the wave length of the mean D line is 5.893×10^{-5} cm. For practical purposes, however, one unit of each kind is not sufficient, and submultiples or multiples of the unit are used as subsidiary units. It is most convenient, as in the metric system, to use only powers of ten as the multipliers, the very various multipliers used in the English system severely tax the memory, and render arithmetic unnecessarily laborious. The choice of ten depends upon the universal use of it as the basis of numeration; if everything could be commenced afresh, probably two would be found more convenient than ten. Even now, for some purposes, such as weighing and interposing resistance coils, it is convenient to reduce to the scale of two.[3]

In some cases it is necessary or convenient to make use of provisional units in terms of which the results are expressed, and to wait for the full expression of our results until our provisional has been accurately determined in terms of the fundamental units.

Thus physicists refer their densities to water, waiting till the number of grams in a cubic centimetre of water has been exactly determined. Chemists refer their atomic weights to oxygen $O = 16$, waiting till the true value of O is known.

Astronomers measure distances in terms of the mean distance of the earth from the sun, waiting till this distance is known in miles.

REFERENCES AND NOTES.

1. Guillaume: *Unités et Etalons.* Paris, 1893. A good account of modern fundamental and derived units.

Chaney: *Our Weights and Measures.* An authoritative account of the English system.

Latimer Clark: *A Dictionary of Metric Measures.* Gives the numerics and their logarithms, which connect a very large number of units.

2. Maxwell: "Dimensions," *Ency. Brit.*, new edition.

S. Lupton: *Numerical Tables and Constants in Elementary Science.*

Everett: *Units and Physical Constants.*

Macfarlane: *Physical Arithmetic*, goes fully into the convertion of ordinary units with problems.

3. Maxwell: *Elementary Treatise on Electricity*, p. 180. A number is expressed in the scale of two by the remainders of successive divisions by two, *e.g.*,

```
2 ) 12
2 ) 6    0     Twelve in the scale of ten is expressed by 1100 in the
2 ) 3    0     scale of two. By subtracting successively the decimal
     1   1     corresponding to $2^{-n}$ it is easy to reduce a common
               decimal to one expressed in the scale of two, thus,
```

.53125 is .10001 or 17/32.

CHAPTER VIII.

AVERAGES.

IN popular language the words average[1] and mean are used as though they were synonymous to express a value intermediate between various measurements of a quantity.

Though the same processes are used, and the results arrived at· are identical in each case, the words express ideas which are in reality quite distinct, they therefore should be carefully defined and accurately used.

If a number of individuals is sufficiently large to form a class, and a quality common to all be measured, the numerics found differ more or less from one another. For simplicity of conception and expression the quality of the class may often be represented with sufficient accuracy for certain purposes by a value intermediate between the values in each individual case. This intermediate value is called an average.

An average then is a value, which may or may

not represent an actual individual, but which expresses the value of a quality in the case of a class.

A mean[2] (*q.v.*), on the other hand, expresses the most probable value of one actual quantity.

An average may be obtained in several different ways, but three methods are most important and most generally used. The methods nearly coincide in their results if the number of instances be large; the first is in general most accurate, but the two latter are often more simple in the case of a great number of instances. In every case all the measurements must be expressed in terms of the same unit.

(i.) The numerics expressing the quality in the case of each individual are added together. The sum divided by the number of individuals gives the arithmetical average,[3] often called simply the average,

$$y = \frac{x_1 + x_2 + \ldots + x_n}{n}.$$

If the number of grains of sulphur per 100 cubic feet of coal gas be determined every hour, and numerics varying from 14 to 16 be obtained, on adding all the results together and dividing by 24 the average sulphur impurity in the gas for the day is obtained.

Occasionally, and especially in statistical results which obey the compound interest law, it is preferable to use the geometrical, which is always smaller than the arithmetical average. To obtain it the

AVERAGES. 43

logarithms of the individual numerics are added together, the sum divided by the number of individuals gives the logarithm of the geometrical average,

$$\log y = \frac{\log x_1 + \log x_2 + \ldots + \log x_n}{n}.$$

If there are only two values of x, the geometrical average is the square root of their product.

The population of England (millions) was in

1801, -	8.9	1841, -	15.9	1871, -	22.7
1811, -	10.2	1851, -	17.9	1881, -	26
1821, -	12	1861, -	20	1891, -	29
1831, -	13.9				

The arithmetical average of these numbers is 17.65, but the geometrical average 16.44 is in all probability a more correct estimate of the population in 1846.

(ii.) If all the individuals are divided into groups so that in each group the individuals are very nearly identical in respect to a quality, the measurement of that quality in the most numerous group may be taken as the average. It of course comes to the same thing, when all the individuals have been measured, to select the numeric which most frequently occurs as the average. It is convenient to follow Prof. Karl Pearson[4] in calling an average obtained by this method—the mode.

After measuring a regiment of guards, if we find 5′ 11″ occur more frequently than another value, we call it the mode of the regiment.

(iii.) Another average is very convenient in some cases. If all the individual results are ranged in the order of their magnitude, the middle result of the row constitutes the median.[5] If all the results are divided into four equally numerous groups or quartiles, the median lies between the two middle groups, the other extremes of which give limits between which it is equally probable that any given result does or does not lie. The median then is a value such that an equal number of the observed values lie on each side of it. It is easily found, unaffected by exceptional cases, and, since the results cluster thickly round it, any error in the estimation of it is usually small and of little moment.

To obtain an average of any real value, the results from which it is derived must be sufficiently numerous and sufficiently widely distributed to be fairly representative of the entire class under discussion. Neglect of this point is a common source of error in statistics, and therefore in the arguments based upon them.

In trade analyses great care must be exercised that the sample operated on is a fair average of the total bulk.

In obtaining the average height of Englishmen we must not select our instances solely from the slums of London, or the dales of Westmoreland.

REFERENCES AND NOTES.

1. The word average is derived from the low Latin *averagium*, which is connected with *averium* and *habere*. *Averagium* meant the transport service of oxen, horses, and carts due from a tenant to his lord, and when the actual service was commuted for a payment, the tenant's proportion of the cess. In Scotland, cart horses are still called cart-avers.

2. The difference between an average and a mean has been well put by Herschel (*Essays*, 404): "An average may exist of the most different objects, as of the heights of houses in a town, or the sizes of books in a library. It may be convenient to convey a general notion of the things averaged, but involves no conception of a natural and recognizable central magnitude, all differences from which ought to be regarded as deviations from a standard. The notion of a mean, on the other hand, does imply such a conception, standing distinguished from an average by this very feature, viz., the regular march of the groups increasing to a maximum and thence again diminishing."

3. The arithmetical average is called by Stanley Jevons the "fictitious mean," to distinguish it from the true mean, which he calls the "precise mean result."

4. *Essays.* "The Chances of Death," p. 11.

5. Mentioned by Laplace: *Mech. Cel.*, iii. 39, and *Essai Philosophique*, Paris, 1840, p. 103; discussed by F. Galton, *Phil. Mag.*, 1875, 33.

CHAPTER IX.

DIFFERENCES.

WHEN two quantities vary simultaneously, a number of values of a variant or function for different values of the variable or argument is frequently presented in the form of a table. The values may be obtained in pure mathematics by calculation from a formula, or in natural science as the result of observations.

By finding a series of differences between the consecutive values of the function given, great assistance may be rendered in testing the accuracy of the figures, in determining the value of the function for values of the argument intermediate between those given in the table, and in finding the form of the function best suited to express the experimental results.

Differences are obtained by subtracting each value of the function from the following one which gives the column of first differences, subtracting each first difference from the succeeding one to obtain the column of second differences, and continuing the

DIFFERENCES.

process until the differences become irregular, or very small.

Several different systems for the notation of differences are in use. It is perhaps most simple to denote values of the variable by x_{-2}, x_{-1}, x_0, x_1, x_2, etc., and the corresponding values of the variant by u_{-2}, u_{-1}, u_0, u_1, u_2, etc. It is usual to mark the line where calculation is to begin, or the values immediately below the new value required by the suffix 0, and values further up the table by negative suffixes. Differences are denoted by a capital Δ prefixed to the u, but the u may be omitted for simplicity. The column in which the difference stands is marked by the affix 1, 2, 3, etc., and its

x	u	$\Delta^1 u$	$\Delta^2 u$	$\Delta^3 u$	$\Delta^4 u$	$\Delta^5 u$
x_{-3}	u_{-3}					
		$\Delta^1_{-\frac{5}{2}}$				
x_{-2}	u_{-2}		Δ^2_{-2}			
		$\Delta^1_{-\frac{3}{2}}$		$\Delta^3_{-\frac{3}{2}}$		
x_{-1}	u_{-1}		Δ^2_{-1}		Δ^4_{-1}	
		$\Delta^1_{-\frac{1}{2}}$		$\Delta^3_{-\frac{1}{2}}$		$\Delta^5_{-\frac{1}{2}}$
x_0	u_0		$\Delta^2_0\,\Delta'_{}$		Δ^4_0	
		$\Delta^1_{\frac{1}{2}}\,\Delta'_0$		$\Delta^3_{\frac{1}{2}}\,\Delta'_{}$		$\Delta^5_{\frac{1}{2}}$
x_1	u_1		$\Delta^2_1\,\Delta^2_0$		$\Delta^4_1\,\Delta^4_{}$	
		$\Delta^1_{\frac{3}{2}}$		$\Delta^3_{\frac{3}{2}}\,\Delta^3_0$		$\Delta^5_{\frac{3}{2}}$
x_2	u_2		Δ^2_2		$\Delta^4_2\,\Delta^4_{}$	
		$\Delta^1_{\frac{5}{2}}$		$\Delta^3_{\frac{5}{2}}$		$\Delta^5_{\frac{5}{2}}\,\Delta^5_{}$
x_3	u_3		Δ^2_3		Δ^4_3	
		$\Delta^1_{\frac{7}{2}}$		$\Delta^3_{\frac{7}{2}}$		
x_4	u_4		Δ^2_4			
		$\Delta^1_{\frac{9}{2}}$				
x_5	u_5					

place in the column by a suffix which may be a whole number or a proper or improper fraction with 2 as the denominator. Differences in the same horizontal line are marked by the same suffixes, and the suffixes in each column are the means between those of the values from which they are derived.

Each column of differences is less by one member than the column to the left of it.

These points will be seen in the table on the preceding page.

Let x represent a series of numbers, and u the corresponding natural logarithm multiplied by 10,000 to avoid decimals:

x	u	Δ^1	Δ^2	Δ^3	Δ^4
1.1	0953				
		870			
1.2	1823		−69		
		801		9	
1.3	2624		−60		0
		741		9	
1.4	3365		−51		−3
		690		6	
1.5	4055		−45		0
		645		6	
1.6	4700		−39		−1
		606		5	
1.7	5306		−34		−2
		572		3	
1.8	5878		−31		
		541			
1.9	6419				

The numbers decrease as we descend each column, and also as we pass towards the right. The irregularity of the numbers in the last column shows that the process has been carried further than the accuracy of the original values warrants; this, of course, is owing to the omission of all figures in the logarithms beyond the fourth.

If it be required to find the value of u for a value of x intermediate between 1.5 and 1.6, 1.5 is taken as x_0, u_0 as 4055, $\Delta^1_{\frac{1}{2}}$ as 645, Δ^2_1 as -39, and $\Delta^3_{\frac{3}{2}}$ as 5. The fourth difference is too uncertain to be available.

Suppose that by accident or the omission of figures any value of the function is in error by $\frac{1}{2} = .5$, this error increases in each column of differences; there will be two errors of .5 in Δ^1, two errors of .5 and one error of 1 in Δ^2, and so on. If we write $s = \frac{n}{2}$ or $\frac{n-1}{2}$, whichever is a whole number, the maximum value for the error in the nth order of differences will in this case be

$$\frac{1}{2} \frac{\overline{n} \cdot \overline{n-1} \cdot \overline{n-2} \ldots \overline{n-s+1}}{1 \cdot 2 \cdot 3 \ldots s}, \text{ or } \frac{1}{2} \frac{\lfloor n}{\lfloor s^2}.$$

This being the average case, if none of the differences exceed twice the above limit, the given values may be assumed to follow a regular law, or to be correct to a unit in the last figure. The quotient of a difference in the last column by twice the above value may be considered to show the maximum error with which the number opposite to it is affected.

For the fourth difference $s = \frac{4}{2} = 2$, and the value of $\frac{4 \cdot 3}{1 \cdot 2}$ is 6, hence the values given in the table are probably correct to the last figure.

If, however, 4702 had been written by mistake for 4700, the differences would have been

```
4055           −43
       647            0
4702           −43           11
       604            11
5306           −32
```

Hence there is an error in the value of 4702, and its probable amount is $\frac{11}{6}$, or about 2.

REFERENCES AND NOTES.

Newcomb: *Logarithmic and other Mathematical Tables*, New York, 1895.

CHAPTER X.

INTERPOLATION.

WHEN a series of values of a function is given for different values of the argument, the process of finding one or more values of the function corresponding to intermediate values of the argument is spoken of as interpolation. A similar process applied beyond the series of given values is called extrapolation; but this process must always be used with caution, and not applied far beyond the limits of the given values.

The problem is treated differently according to whether the given values of the argument are equidistant as in an ordinary table of logarithms or not, as is generally the case in observational results.

If the differences between the given values of the function are comparatively small and regular, it is generally sufficiently accurate to assume that the change in the value of the function is proportional to the change in the value of the variable. The calculation of the proportional parts is obviated by the use of small tables, or by the assistance of a slide rule. If the intervals are large and the first differences increase or decrease with comparative rapidity, the method of simple proportion becomes so inaccurate as to be useless.

The general form of the problem is, given that a series of equidistant values of the argument $x_{-3} x_{-2} x_{-1} x_0 x_1 x_2 x_3$, etc., make the value of the function $u_{-3} u_{-2} u_{-1} u_0 u_1 u_2 u_3$, etc., to find the value of the function u_n where n is a proper fraction lying between x_0 and x_1.

Making the assumption that the formula connecting the argument and function is a continuous one, or that there is no sudden change of value between the given values of the function, Newton's formula of interpolation gives

$$u_n = u_0 + \frac{n}{1}\Delta^1{}_{\frac{1}{2}} + \frac{n \cdot \overline{n-1}}{1 \cdot 2}\Delta^2{}_1 + \frac{n \cdot \overline{n-1} \cdot \overline{n-2}}{1 \cdot 2 \cdot 3}\Delta^3{}_{\frac{3}{2}} + \text{etc.},$$

the series being continued until the differences become negligible. Since n is less than 1 the coefficients of the even terms are negative.

When u_0 is not the first term given and the fifth differences are approximately equal, it is more convenient to use Bessel's formula:

$$u_n = u_0 + \frac{n}{1}\Delta^1{}_{\frac{1}{2}} + \frac{n \cdot \overline{n-1}}{1 \cdot 2} \frac{\Delta^2{}_0 + \Delta^2{}_1}{2} + \frac{n \cdot \overline{n-1} \cdot \overline{n-\frac{1}{2}}}{1 \cdot 2 \cdot 3}\Delta^3{}_{\frac{1}{2}}$$

$$+ \frac{\overline{n+1} \cdot n \cdot \overline{n-1} \cdot \overline{n-2}}{1 \cdot 2 \cdot 3 \cdot 4} \frac{\Delta^4{}_0 + \Delta^4{}_1}{2}$$

$$+ \frac{\overline{n+1} \cdot n \cdot \overline{n-1} \cdot \overline{n-2} \cdot \overline{n-\frac{1}{2}}}{1 \cdot 2 \cdot 3 \cdot 4 \cdot 5}\Delta^5{}_{\frac{1}{2}}.$$

INTERPOLATION.

Since n is less than one, the coefficient of Δ^2 is always negative; the coefficient of either Δ^5 or Δ^3 is negative according to whether n is $<$ or $>.5$; if $n=.5$, Δ^3 and Δ^5 vanish.

Suppose that it is required from the following table to find the value of $e^{.55}$, which is 1.733252, etc. The fifth differences are so irregular and small that they may be neglected.

x	$u=e^{\frac{x}{10}}$	Δ^1	Δ^2	Δ^3	Δ^4	Δ^5
1	1.10517					
		.11623				
2	1.22140		.01223			
		.12846		.00128		
3	1.34986		.01351		.00013	
		.14197		.00141		+.00005
4	1.49183		.01492		.00018	
		.15689		.00159		−.00005
5	1.64872		.01651		.00013	
		.17340		.00172		+.00008
6	1.82212		.01823		.00021	
		.19163		.00193		−.00003
7	2.01375		.02016		.00018	
		.21179		.00211		
8	2.22554		.02227			
		.23406				
9	2.45960					

By simple proportion
$$u_0 = 1.64872$$
$$\tfrac{1}{2}\Delta^1{}_{\tfrac{1}{2}} = .08670$$
$$u_n = 1.73542$$

in which the fourth figure is incorrect.

By Newton's method—

$u_0 = 1.64872$
$\tfrac{1}{2}\Delta^1_{\frac{1}{2}} = .08670$
$\tfrac{1}{8}\Delta^2_1 = -.00228$
$.0625\,\Delta^3_{\frac{3}{2}} = .00012$
$.039\,\Delta^4_2 = -.000007$
$u_n = 1.73325$

By Bessel's method—

$u_0 = 1.64872$
$\tfrac{1}{2}\Delta^1_{\frac{1}{2}} = .08670$
$\tfrac{1}{8}\dfrac{\Delta^2_0 + \Delta^2_1}{2} = -.00217$
$.02344\,\dfrac{\Delta^4_0 + \Delta^4_1}{2} = .000004$
$u_n = 1.73325$

In this case both methods give the results correct to the last figure, but Bessel's method requires fewer figures than Newton's.

The more general and more difficult problem is, given a series of values for a rational integral function which are not equidistant to find an intermediate value.

The method generally used is known by the name of Lagrange.[1] Let u become $u_0\,u_1\,u_2\,u_3$ when x becomes $x_0\,x_1\,x_2\,x_3$ given a value x find u_x.

$$u_x = \frac{(x-x_1)(x-x_2)\ldots(x-x_n)}{(x_0-x_1)(x_0-x_2)\ldots(x_0-x_n)}u_0$$
$$+ \frac{(x-x_0)(x-x_2)\ldots(x-x_n)}{(x_1-x_0)(x_1-x_2)\ldots(x_1-x_n)}u_1$$
$$+ \frac{(x-x_0)(x-x_1)(x-x_3)\ldots(x-x_n)}{(x_2-x_0)(x_2-x_1)(x_2-x_1)\ldots(x_2-x_n)}u_2 + \text{etc.}$$

If the function be periodic we may assume

$$u_x = \frac{\sin(x-x_1)/2 \times \sin(x-x_2)/2 \times \ldots \times \sin(x-x_n)/2}{\sin(x_0-x_1)/2 \times \sin(x_0-x_2)/2 \times \ldots \times \sin(x_0-x_n)/2}u_0$$
$$+ \text{etc.}$$

INTERPOLATION.

A very simple and convenient formula to convert readings on the scale of a spectroscope into wavelengths may easily be arrived at. In the case of substances of medium refractive index the increment of the index of refraction is nearly proportional to the increment of the square of the reciprocal of the wave length.[2] Hence if the indices of refraction, the angular deviation, or the scale readings of three lines near together are given, and also the wave lengths of two of them, the wave length of the third can be calculated.

Suppose λ_0, λ_1, λ_2 the wave lengths n_0, n_1, n_2 the scale readings of the three lines substituting in Lagrange's formula.

$$\frac{1}{\lambda_1^2} = \frac{1}{\lambda_0^2} \frac{n_1 - n_2}{n_0 - n_2} + \frac{1}{\lambda_2^2} \frac{n_1 - n_0}{n_2 - n_0}.$$

In the case of three bright lines of magnesium

$\lambda_0 = 5183$, $\lambda_2 = 5167$, $n_0 = 74.5$, $n_1 = 74.8$, $n_2 = 75$,

$$\therefore \frac{1}{\lambda_1^2} = \frac{1}{(5183)^2} \frac{.2}{.5} + \frac{1}{(5167)^2} \frac{.3}{.5},$$

$\therefore \lambda^2 = 5173$ instead of 5172.

REFERENCES AND NOTES.

1. Boole: *Finite Differences*, 2nd edition, 38, 42.
2. Stokes: *B.A. Report*, 1849.
W. Gibbs: *Silliman's American Journal*, II. l. 45. The result is only approximate.

CHAPTER XI.

MENSURATION.

IT is frequently required to find the length of a curve, the area of a plane, the surface or volume of a solid, or the contents of a vessel. If the equations to the curve, the bounding line, or surface be given, it is in general necessary to consider the line, surface, or volume to be cut up into a number of infinitesimal elements, and then to find the sum of all these elements by integration between the limits imposed by the question.

If the lines are, or can be, broken up into straight lines, or certain well-known curves, and the surfaces are plane or formed by the revolution of well-known curves, the lengths, areas, and volumes can generally be found by the simple formulae given in treatises on Mensuration.[1]

Various mechanical appliances are used to avoid difficult measurements or tedious calculations. The course of a curve may be followed by bending a strip of card along it, and the straightened card may then be directly measured. For many purposes the ofisometer, which consists of a rough-edged wheel rotating on a screw, is very convenient. The wheel is run forward along the curved or broken line, and then backwards through the same number of revolutions along a straight measure.

Many ingenious planimeters have been suggested to find the area of planes by tracing their contours. The one invented by Prof. Amsler[2] is in most general use; by its aid the area of an irregular figure, such as an indicator diagram is readily found.

The volume of a vessel is generally found by weighing the water or mercury which it will hold, and that of a solid by finding the weight it apparently loses when immersed in water or some liquid of known density.

In the absence of a planimeter the following rules may be of use in measuring a plane area.

A convenient line is drawn across or along one side of the figure, and divided into a number of equal parts. From each point ordinates or offsets are drawn at right angles to meet the boundary of the area, and the length of each offset is measured. Speaking generally, the more numerous the offsets the more accurate the results, but great accuracy is not to be expected if there is very great difference between the ordinates, or if the curve meets the base line nearly at a right angle. This difficulty may be often avoided by the proper choice of a base line, or by treating different portions of the area separately.

The most simple rule, which is generally sufficiently accurate, is:—

Add half the sum of the first and last ordinates to the sum of all the other ordinates, and multiply by the common distance between the ordinates.

It is generally more simple to obtain greater accuracy by taking more ordinates than by applying a more complicated rule, such as that of Weddle.[3]

Divide the base into six equal parts, add into one sum the first and every alternate ordinate, five times the sum of the intermediate ordinates, and the middle ordinate. Multiply by $\frac{3}{10}$ of the common distance between the ordinates.

To find the area of a portion of a semicircle, radius 6, cut off by the curve, the diameter, and ordinates at $x = \pm 3$.

$$y = \sqrt{36 - x^2}.$$

$x = -3, \quad y_1 = 5.19615;$ $\quad x = \frac{1}{2}, \quad y = 5.97913;$
$x = -2\frac{1}{2}, \quad y = 5.45436;$ $\quad x = 1, \quad y_5 = 5.91608;$
$x = -2, \quad y_2 = 5.65685;$ $\quad x = 1\frac{1}{2}, \quad y = 5.80947;$
$x = -1\frac{1}{2}, \quad y = 5.80947;$ $\quad x = 2, \quad y_6 = 5.65685;$
$x = -1, \quad y_3 = 5.91608;$ $\quad x = 2\frac{1}{2}, \quad y = 5.45436;$
$x = -\frac{1}{2}, \quad y = 5.97913;$ $\quad x = 3, \quad y_7 = 5.19615.$
$x = 0, \quad y_4 = 6.00000;$

Take first only the seven ordinates with subscript figures which are one inch apart. Add one extreme to the other values, the area is found to be 34.342. Taking the whole thirteen ordinates half an inch

MENSURATION. 59

apart, add them all with the exception of one extreme and multiply by $\frac{1}{2}$, the area is found to be 34.414.
Applying Weddle's rule, add
$$y_1+y_3+y_5+y_7+5(y_2+y_4+y_6)+y_4,$$
and multiply by $\frac{3}{10}$, the area is found to be 34.4379.
Since the figure is made up of two right-angled triangles and a circular sector containing 60°, its area is
$$3 \times 5.19615 + \frac{36\pi \times 60}{360} = 15.58845 + 18.849556 = 34.438.$$
Hence the accuracy of the first method increases with the number of ordinates, and in this case Weddle's rule is nearly exact.

REFERENCES AND NOTES.

1. Lupton, *Numerical Tables*, p. 7.
2. Williamson, *Integral Calculus*, 214, for theory.
3. Boole, *A Treatise on the Calculus of Finite Differences*, p. 47. Many formulae have been proposed by Stirling, Simpson, and others, that of Weddle combines simplicity with considerable accuracy.

CHAPTER XII.

THE USE OF TABLES.

IN pure mathematics the results of tedious arithmetical operations, which are frequently required, are registered in tables, which facilitate and conduce to the accuracy of calculation. If every logarithm had to be calculated when required, the labour would entirely prohibit their use; but logarithms, when printed in tables, are of the very greatest assistance to the calculator.

The most generally useful tables are multiples, Crelle's Rechentafeln go to 1000 × 1000; reciprocals, Oakes, gives five figure numbers to seven figures; squares, cubes, square and cube roots, Barlow gives to 10,000 to seven figures; logarithms, of numbers and of the trigonometrical functions.[1]

It is useless, misleading, and tiresome to carry the calculation far beyond the accuracy of the observation, hence, for the ordinary work of a beginner, Barlow and Schlömilch's logarithms are amply sufficient. If due care be exercised, they will

THE USE OF TABLES. 61

give an accuracy of one or two units in the fifth place.

The series of values of the variable, by which the table is arranged, is called the argument, and the corresponding values of the variant the entry. In general, each entry corresponds to only one argument, but in a few cases such as the multiplication table, there are two arguments to each entry. Tables with two or more arguments are liable to become very complicated and bulky.

Tables are most convenient when the differences between consecutive arguments and entries are so small, that intermediate values may be found by simple proportion; if this is not the case, second, and even third differences must be used.

$$u_n = u_0 + \frac{n}{1}\Delta^1 + \frac{n \cdot \overline{n-1}}{1 \cdot 2}\Delta^2.$$

Thus, to find log 108.79379, having given

log 108 = .03342376
       .00400274
109 = .03742650   −.00003655.
       .00396619
110 = .04139269

u_0 = .03342376

$\frac{n}{1}\Delta^1$ = .00317733

n = .79379

$\frac{n \cdot \overline{n-1}}{1 \cdot 2}\Delta^2$ = .00000299

    2.03660408

The more correct value is 2.036604 1063326....

In many cases special formulae for the differences are more convenient and exact than the general one. Thus,

$$\log \overline{a \pm x} = \log a \pm 2M\left(\frac{x}{2a \pm x} + \frac{1}{3}\frac{x^3}{(2a \pm x)^3} + \cdots\right)$$

where the tabular argument is represented by a and the difference, or if it be greater than 0.5, the complement of the difference is represented by $\pm x$. In the case given above,

$$2a = 218, \quad 2a - x = 217.79379, \quad x = -.20621,$$

$$\log a = 2.03742650,$$

$$\frac{2Mx}{2a - x} = -.00082239$$

$$\overline{}$$

$$2.03660411$$

The error is < 4 in the ninth decimal place, the omitted term is less than 3 in the ninth place.

The sin, cos, tan of an angle which is nearly 0° or 90° approach their limiting values 0, 1, ∞, and the differences, especially of the logarithms of the functions, become too irregular to be available. Thus the differences for 1′ of a 5-figure logarithmic table are

	sin	cos	tan		sin	cos	tan
1°	717	0	718	85°	2	145	146
3°	240	0	241	87°	1	242	243
5°	144	1	145	89°	0	630	730

If possible it is generally best to modify the experiments so as to avoid very small or very large

THE USE OF TABLES.

angles; if this is impossible, special methods of calculation must be used.

Some tables give a much closer range of arguments for the first few degrees, and almost all give Delambre's method or Maskelyne's modification of it.

The following series due to Borda are not often given, but are easily applied in the case of small angles. Convert the angle into circular measure.

$\log \sin \theta$

$$= 10 + \log \theta - M\left\{\frac{\theta^2}{6} + \frac{\theta^4}{180} + \frac{\theta^6}{2835} + \frac{\theta^8}{37800} + \text{etc.}\right\},$$

$\log \cos \theta$

$$= 10 - M\left\{\frac{\theta^2}{2} + \frac{\theta^4}{12} + \frac{\theta^6}{45} + \frac{17\theta^8}{2520} + \frac{31\theta^{10}}{14175} + \text{etc.}\right\},$$

$\log \tan \theta$

$$= 10 + \log \theta + M\left\{\frac{\theta^2}{3} + \frac{7\theta^4}{90} + \frac{62\theta^6}{2835} + \frac{127\theta^8}{18900} + \text{etc.}\right\}.$$

These series decrease very rapidly for small values of θ and hold good at least from 0° to 90°.

Thus in the case of $\theta = 5° \ 43' \ 46''.48 = .1$ radian.

$$\log \sin \theta = 10 - 1 - M\left(\frac{.01}{6} + \frac{.0001}{180} + \frac{.000001}{2835} + \frac{.00000001}{37800}\right)$$

$$= 9 - M(.00166\ 72225\ 75221)$$

$$= 9 - 0.00072\ 40655\ 64522$$

$$= 8.99927\ 59344\ 35477.$$

The fourth term of the series only affects the thirteenth decimal place, and the third term the tenth, so that a few terms of the series suffice for small values of θ.

In large tables second differences may be required to obtain correct results with all the trigonometrical functions. Suppose the argument be $10''$ or $1''$,

$$\log \sin \overline{a+x} = \log \sin a + x\Delta^1/10 + \tfrac{1}{2}x(10-x)\Delta^2/100,$$
$$\log \cos \overline{a+x} = \log \cos a - x\Delta^1/10 + \tfrac{1}{2}x(10-x)\Delta^2/100,$$
$$\log \tan \overline{a+x} = \log \tan a + x\Delta^1/10 \pm \tfrac{1}{2}x(10-x)\Delta^2/100,$$

where in the last term $+$ is used for angles less than $45°$, and $-$ for angles greater than $45°$.

Almost every branch of science is supplied with a variety of special tables[2]; a few of those most generally useful in physics and chemistry will be mentioned in the notes, and the methods by which they are calculated from experimental results will be discussed later.

REFERENCES AND NOTES.

1. De Morgan: "Tables" in *English Encyclopædia.*
B. A.: *Report of Committee on Mathematical Tables,* 1873-1875, gives very complete critical lists of tables in pure mathematics.

A few more recent or specially convenient tables of logarithms may be mentioned:

Bottomley: 4-figure mathematical tables to 6'.
Höuel ⎱ 5-figure tables to 1'.
Schlömilch ⎰
Jones ⎱ 6-figure tables to 1'.
Lodge ⎰
Chambers: 7-figure tables to 1'.

THE USE OF TABLES.

Bruhns } 7-figure tables to 10".
Schrön

Arithmometers are now generally used instead of more powerful tables.

2. Lupton : *Numerical Tables and Constants in Elementary Science.*
Everett : *Units and Physical Constants.*
Agenda du Chemiste.
Annuaire publié par le Bureau des Longitudes.
Biedermann : *Chemiker-Kalender.*
Hospitalier : *Formulaire de l'Electricien.*
Gray : *Smithsonian Physical Tables.*
Landolt and Börnstein : *Physikalisch Chemische Tabellen.*

CHAPTER XIII.

ERRORS.

WHEN a quantity is measured with all possible accuracy many times in succession, the numerics expressing the results are found to differ by amounts, which are generally small, but occasionally considerable in comparison with the quantity measured.

Though these differences may be decreased by improved methods, better instruments, or greater skill they can never be entirely removed. Apparent agreement merely shows that the methods are in fault, or that the measures have not been carried to the fullest possible extent.

Incorrect results may be obtained owing to defects in the method, in the instrument, or in the observer. Thus in volumetric analysis the end of the reaction may not be sufficiently sharp, the burette may not be accurately graduated, the observer may not keep his line of sight horizontal in reading the meniscus.

All these sources of error can be decreased by due precautions; a different reaction, or a better

ERRORS. 67

indicator, may be used, the burette may be calibrated, the eye may be assisted by a horizontal telescope.

Such sources of error as these, when so large that taken singly, they can be reduced by greater care, are spoken of as constant errors; though the term is misleading since they are generally not exactly the same in two observations.

All observers also are liable to make mistakes, such as misreading a figure on the burette, or forgetting to add in a weight. Mistakes are often corrected by the check of a second observer.

When all mistakes have been removed, and all constant errors have been reduced as far as possible, the small differences between the numerics found are supposed to be due to, and are themselves called errors of observation. These errors are supposed to be in each case the resultant of a number of partial errors, to which every separate portion of each observation is liable.

The light may vary in intensity, or the eye of the observer in precision, so that he cannot accurately distinguish a change of colour or the first formation of a precipitate. The burette may vary in temperature or may drain more or less thoroughly. The eye of the observer may not be exactly in the optical axis of the telescope, or the position of the screen may affect the reading of the meniscus. All these and many other partial errors, each of which may vary within narrow limits, combine to produce the total error of the observation. In any given case the partial errors are very numerous, and, if we assume that the great majority of them are equally likely to err in excess or defect of the true value, it is about an equal chance that their sum, the total error of the observation, is positive or negative. It is generally assumed, that, when any quantity is measured many times in succession with the greatest possible accuracy, positive and negative errors are about equally numerous, and that the sum of the positive is about equal to the sum of the negative errors.

Again, if it be assumed that each partial error is nearly constant and is equally likely to be positive or negative, the most unlikely case is that all the partial errors should be positive or all negative, and the most likely case that half should be positive and half negative.[1] Hence in any series of results, the smaller the errors the more likely and the more frequent the cases.

Since, in any carefully conducted experiment, the partial errors are extremely small, and their number though large is not infinite, except in rare instances, their sum is very small in comparison with the quantity measured. It is therefore far more advantageous to give time and attention to the removal of mistakes and constant errors, than to take excessive trouble in attempting to allow for errors by calculation.[2]

If the positive and negative partial errors happen to balance one another, the true result is obtained, but we have no means of knowing when this is the case. Hence we have no means of arriving at the actual value, of which our observational results are more or less correct representatives.

It is generally assumed that a mean of the observational results gives the best and most probable representative of the actual value, and that we can render the mean more accurate and more probable by increasing the number of the observations.

The above remarks are very theoretical, and are merely suggestive of what really takes place in any observation. The best proof of the whole theory of errors is that the results obtained by the use of it are found approximately to agree with the results of long series of accurate observations.

As the remarks in this section are very important, somewhat vague and difficult to remember, it may be worth while to recapitulate them briefly.

All observations are liable to mistakes due to the observer and to constant errors due to the method, the instrument, or the observer. These may be of any magnitude compared to the quantity measured, and can be removed or decreased by greater care and skill.

When a considerable number of observations are made with the greatest care under similar conditions upon the same quantity, and all mistakes and constant errors have been as far as possible removed, the differences between the results are due to, and are themselves called, errors of observation.

(i.) Errors in excess and in defect of the true value

are about equally numerous, and the sums of the positive and of the negative errors are about equal.

(ii.) Errors are small in comparison with the quantity measured, and are more numerous the smaller they are.

REFERENCES AND NOTES.

1. The chance of the occurrence of any special combination of n partial errors, each of which is equally likely to be positive or negative, is expressed by the value of the $(r+1)^{th}$ term in the expansion of $\left(\dfrac{1}{2}+\dfrac{1}{2}\right)^n$, where r varies from 0 to n, this is

$$\frac{\lfloor n}{\lfloor n-r \lfloor r} \left(\frac{1}{2}\right)^{n-r} \times \left(\frac{1}{2}\right)^r.$$

If the chances of the occurrence of positive and negative partial errors is not equal; $\frac{1}{2}$ must be replaced by the requisite fractions. To find the greatest term in the expansion of the binomial $(x+y)^n$, find

$$s = \frac{(n+1)y}{x+y}.$$

If s is a whole number there are two equal maximum terms, the sth and the $(s+1)^{th}$, if s be a mixed number neglect the fractional part, and the $(s+1)^{th}$ is the maximum term. The maximum term expresses the most frequent and most probable case.

If there are 100 partial errors, the maximum term is the fifty-first, the chance that all are positive or all negative is $1/1.27 \times 10^{30}$; the chance that one error is positive or negative and the rest negative or positive is $100/1.27 \times 10^{30}$; the chance that half are positive and half negative $10^{29}/1.27 \times 10^{30}$, or about $1/13$; and these chances represent the average occurrence of each combination in a large number of instances.

2. Bertrand: *Probabilités*, Introd., xxxix.

CHAPTER XIV.

MEANS.

THE word means[1] is used in mathematics to signify a term or a series of terms inserted according to a given rule between two given extreme terms. The eleven different kinds of means, which were recognized in the middle ages, have only an antiquarian interest.

The use of the term in science is entirely different. When one observation only has been made, the result must be taken as the most probable value of the quantity measured, since no further data are available. If several results have been obtained, which differ slightly owing to errors of observation, it is generally assumed that the most advantageous and most probable approximation to the true value can be found by taking a mean of the results. In scientific language then, the word mean signifies a value derived from a series of observational results, and intermediate between them, which is believed to express the most probable and most advantageous value of the quantity measured.

Many varieties of means are conceivable, and the choice in any special case depends upon several considerations.

Thus we may range the results obtained in the order of their magnitude and take the middle term, the median, as the most probable mean. We may count the number of results lying between equal narrow limits, and assume that the value which occurs most frequently, the mode, is the most probable mean. We may assume that the sum of the positive is equal to the sum of the negative errors, and take the arithmetical mean of the observational results by adding them all together and dividing by the number of them. This mean is far the most commonly used, and is generally spoken of as *the* mean.

Some mathematicians regard the assumption that the arithmetical mean is the best and most probable value to choose as axiomatic; others, with Encke, have attempted to prove it, but all the proposed proofs are more or less unsatisfactory. The most advantageous and the most probable mean need not necessarily coincide, and in each case a different "law of error" gives a different result.

If the usual form of the law of error holds good and the observations are sufficiently numerous, the median, the mode, and the mean coincide, and the last is generally adopted for convenience of calculation; but the least unsatisfactory proofs of the law of error depend upon the assumption that the arithmetical is the most probable mean.[2]

MEANS.

If we assume with Laplace, that the mean, which makes the sum of the squares of the errors least, is most advantageous, we again arrive at the arithmetical mean.[3]

As we are not justified in assuming that the mean of a series of observational results is exactly equal to the actual quantity measured, we cannot in strictness call the difference between the mean adopted and any observational result the error of the observation; the word residual is generally used to express this difference. Of course, if the mean is assumed to be equal to the true value, the residuals become equal to the errors.

REFERENCES AND NOTES.

1. De Morgan: "Mean" in *English Encyclopædia*. The word mean is derived from the Old French *meien*, a contracted form of the Latin *medianus*, from *medius*.
2. Cf. Airy: *Theory of Errors*, p. 53.
3. Cf. Chauvenet: *Method of Least Squares*, p. 476. The common method by differentiating is not quite satisfactory.

CHAPTER XV.

THE LAW OF ERROR.

It is a matter of much interest and importance to find a formula which will connect the number of errors of any given magnitude with that magnitude, or if y expresses the number of errors equal to $\pm x$ to determine the form of the function $y = \phi(x)$[1].

The general problem of connecting the frequency and magnitude of occurrences in any given case by an equation or curve is very difficult, and can only be solved approximately,[2] but in the case of errors of observation the problem is simplified by the following considerations. Positive and negative errors are found to be about equally numerous, x must therefore occur only in even powers; comparatively large errors occur very rarely, hence y must decrease very rapidly as x increases.

Several different lines of argument, no one of which is quite free from objection, lead to the equation $y = e^{-x^2/c^2}$, where e is the Naperian base 2.718..., and c is a numerical value of the same kind as x, which

THE LAW OF ERROR.

is known as the modulus.[3] If h be written for $1/c$, the equation is conveniently expressed as

$$y = \exp.(-h^2 x^2),$$

and h is called the measure of precision of the observations, since the more accurate they are the greater is h. In any given case c or h can generally be put equal to one, or omitted, if it be remembered that all measurements and values are given in terms of them.

The value of y corresponding to any value of hx is easily found, and from these values the graph can be traced.[4]

hx	0	.5	1	1.5	2	2.5	3	5
y	1	.7788	.3679	.1054	.0183	.0019	.0001	$.0^{10}1$

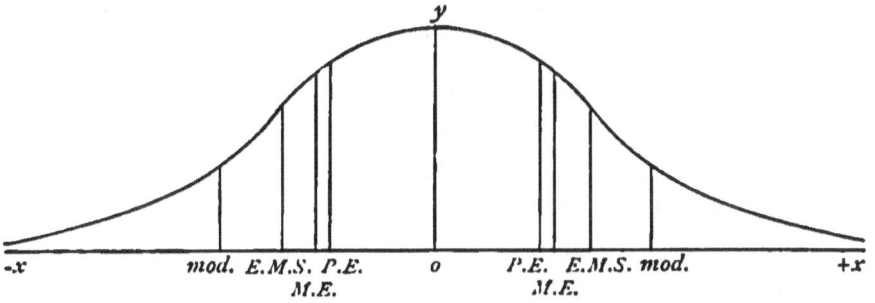

$-x$ mod. E.M.S. P.E.   0   P.E. E.M.S. mod. $+x$
            M.E.   M.E.

The ordinate represents the frequency of an error, the magnitude of which is represented by the corresponding abscissa. The curve is symmetrical about the y-axis, and shows points of inflection at

$$x = \pm 1/\sqrt{2} = \pm 0.7071$$

on each side of it; it rapidly approaches, though it never meets, the x-axis. The area of the whole curve is $c\sqrt{\pi} = 1.77245 \times c$. If the base between any two abscissae $+x$ and $-x$ be divided into a number of small parts, each equal to a, and ordinates be drawn at each subdivision, the area contained by the curve, the base, and the two extreme ordinates is nearly·

a {one extreme + all the intermediate ordinates}.
More exact values are obtained by integration.[5]

The probability of the occurrence of an error $< \pm x$ is expressed by the ratio of the area contained between the curve, the two ordinates at $+x$ and $-x$, and the x-axis, to the area of the whole curve. The value of this ratio for different values of x is given in the second column of Table I.

In practice, it is more convenient to replace the modulus by one of the following so-called "errors" obtained by calculation from the residuals.

If all the positive residuals be added together, the sum divided by the number of them gives the mean positive error; the mean negative error may be obtained in the same way from the negative residuals. Half the sum of the mean positive and of the mean negative error gives the mean error;

$$\text{M.E.} = c/\sqrt{\pi} = c \times 0.564189.$$

If the sum of the positive and the sum of the negative residuals do not differ much, as is usually the case when the observations are numerous, all the residuals

THE LAW OF ERROR.

may be added together without regard to sign. The sum divided by the number gives the M.E.

If the squares of the residuals be added together and the sum be divided by the number of observations less one,[6] the square root of the result gives the error of mean square;

$$\text{E.M.S.} = c/\sqrt{2} = c \times 0.707107.$$

The E.M.S. can also be obtained by subtracting n times the square of the mean from the sum of the squares of the observational results, dividing by $n-1$, and taking the square root of the quotient.

The probable error is an error of such a value, that half the results lie on each side of it, so that it is an equal chance if any given error is greater or less than the probable error. It is the value of x which bisects each half of the probability curve.

$$\text{P.E.} = c \times 0.476948.$$

Some writers, especially in America, call the E.M.S. the mean error, omitting all notice of the true M.E. Though it is really immaterial which of these errors is adopted for practical use, generally either the

	c.	M.E.	E.M.S.	P.E.
In terms of c.	1.0000	0.5642	0.7071	0.4769
,, ,, ,, M.E.	1.7725	1.0000	1.2533	0.8454
,, ,, ,, E.M.S.	1.4142	0.7979	1.0000	0.6745
,, ,, ,, P.E.	2.0967	1.1829	1.4826	1.0000

E.M.S. or the P.E. is selected. The modulus, the mean error, the error of mean square, and the probable error, are represented by four positive, and by four negative abscissae, as shown in the figure. They are connected by the numerics shown in the preceding table.

A convenient test of the fidelity of the observations and of the accuracy of the calculations is afforded by the fact, that in any sufficiently numerous series of observations, the mean of the squares of the errors divided by the square of the mean of the errors is constant;
$$\Sigma e^2/n \div (\Sigma e/n)^2 = \pi/2 = 1.57.$$
Thus twenty-four readings of a certain angle in the United States coast survey gave $3.84/2.56 = 1.5$.

Though the safe application of the law of error or of results derived from it requires the consideration of a considerable number of results, it may be sufficiently exemplified by the use of a small number.

·Suppose the following five readings of a barometer have been obtained:

M.M.	v.	v^2.
760.03	+.03	.0009
759.95	−.05	.0025
760.08	+.08	.0064
759.98	−.02	.0004
759.96	−.04	.0016
760.00	.00	.0118

THE LAW OF ERROR.

The mean result is 760 M.M., and the sum of the residuals is nothing. The E.M.S. is

$$\sqrt{.0118/4} = \pm 0.0543.$$

The probable error of any one observation is

$$0.6745 \times \text{E.M.S.}$$

The probable error of the mean of n observations is

$$0.6745 \times \text{E.M.S.}/\sqrt{n} \text{ or } 0.6745 \times 0.0543/\sqrt{5} = \pm 0.0164.$$

Hence it is an equal chance that 760 M.M. does not err in excess or defect of the true value by so much as ±0.0164 M.M.

The square root of the sum of the squares of the residuals multiplied by the proper number from Table II., gives the probable error of any one of, or of the mean of all the observations.

If the abscissa be measured in terms of the probable error, instead of in terms of the modulus, or if t be written for $0.479636 \times x/\text{P.E.}$, the area of the curve cut off by each positive and negative ordinate represents the number of observations, the error of which is smaller than the abscissa. Thus, if $t=1$, half the observations have errors <1. The ratio of the area cut off by a given abscissa to the total area of the curve is given in the third column of Table I.

Airy found the probable error of 636 Greenwich observations of the N.P.D. of Polaris to be $0''.5711$. To find the number of observations with errors $>2''$, take $t = 2''/0''.5711 = 3.5$ nearly. From the table, if $t=3.5$, $T=0.98176$, hence the proportion of results with errors greater than 2″ is 0.01824; or out of 636 observations, 636 × 0.01824 or about 11.6 may be expected to have errors greater than 2″.

Since all the proofs of the law of error are more or less unsatisfactory, a series of reductions of a great number of accurate observations, such as those of Airy on Polaris, and of Bessel on Sirius and Altair are extremely valuable, since they show that the numbers given by theory are in very fair accordance with those actually found. In each case, however, comparatively large errors were found to be rather more frequent than theory warranted. This shows that the extension of the theoretical limits to $\pm \infty$ has not introduced any inaccuracy.

The ordinate through the vertex of the ordinary probability curve may be considered from three points of view:

(i.) Since the ordinate bisects the area enclosed by the curve and the base, the sums of the magnitudes of the errors on each side of it are equal. It is the ordinate "to which the arithmetical mean of all the different values directs us."

(ii.) Since the ordinate bisects the base, an equal number of observational results lie on each side of it, or it corresponds to the median.

(iii.) Since it is the maximum ordinate, the result 'no error' which it represents is more probable than a result with any given error. It has been called "the mode" by Prof. K. Pearson.

If, then, the curve representing the frequency of

errors be symmetrical, as is most generally the case, the mean, the median, and the mode coincide. If, however, the frequency curve be cut short or be skew these three kinds of means may differ more or less widely, and it becomes a matter of doubt and difficulty to select the most advantageous and most accurate one in any given case.

REFERENCES AND NOTES.

1. Mansfield Merriman: *A Text-book of Least Squares.* The practical and theoretical portions are separated, so that the book can be used without much mathematical training. A useful bibliography is given.

Chauvenet: *The Method of Least Squares*; a reprint of the well-known manual of astronomy.

Airy: *The Errors of Observations*, more difficult.

2. Pearson: *Phil. Trans.*, 185, A. 72.

Sheppherd: *On the Geometrical Treatment of the Normal Curve*, etc.; read R.S., Nov. 25th, 1897.

3. Edgeworth: *Trans. Camb. Phil. Soc.*, 1887. xiv. 2. 138, discusses several possible laws of error.

Glaisher: *Mem. Astron. Soc.*, 1871. 2. xxxix. 75, criticises various proofs of the ordinary law of error.

Herschel: *Essays*, 398, gives a simple proof of the law of error, which, however, is far from being universally accepted as satisfactory.

4. $\log e^{-h^2 x^2} = -0.43429 h^2 x^2$.

5. To integrate $\int e^{-x^2/c^2}$, that is to find the area of the probability curve or of a portion of it cut off by two ordinates. If all values are expressed in terms of the modulus $c = 1$. Since the curve is symmetrical

$$\int_{-\infty}^{\infty} e^{-x^2} dx = 2\int_{0}^{\infty} e^{-x^2} dx = 2\int_{0}^{x_1} e^{-x^2} dx + 2\int_{x_1}^{\infty} e^{-x^2} dx.$$

To integrate $\int_{0}^{\infty} e^{-x^2} dx = k$ suppose.

L. N.   F

Write ax for x, and therefore $a\,dx$ for dx,
$$\int_0^\infty e^{-a^2x^2}a\,dx = k.$$

Multiply by e^{-a^2},
$$\int_0^\infty e^{-a^2(1+x^2)}a\,dx = ke^{-a^2},$$

$$\int_0^\infty\int_0^\infty e^{-a^2(1+x^2)}a\,da\,dx = \int_0^\infty ke^{-a^2}da = k^2,$$

but $\int_0^\infty e^{-a^2(1+x^2)}a\,da = \frac{1}{2}\int_0^\infty e^{-a^2(1+x^2)}d(a^2) = \frac{1}{2}\frac{1}{1+x^2}.$

Again, $\int_0^\infty \frac{dx}{1+x^2} = 2k^2 = \left[\tan^{-1}x\right]_0^\infty = \pi/2$;

$$\therefore\ k = \sqrt{\pi}/2\,;$$

$$\therefore\ \int_{-\infty}^\infty e^{-x^2}dx = 2k = \sqrt{\pi} = 1.77245385\ldots\,;$$

also, $2\int_0^\infty e^{-x^2}dx = \frac{2}{2}\Gamma\!\left(\frac{1}{2}\right) = \sqrt{\pi}.$

If x_1 be small write in the exponential value,
$$\int_0^{x_1} e^{-x^2}dx = \int_0^{x_1}\left(1 - \frac{x^2}{1} + \frac{x^4}{1.2} - \frac{x^6}{1.2.3} + \text{etc.}\right)dx,$$

$$\int_0^{x_1} e^{-x^2}dx = \left[x - \frac{1}{2}\frac{x^3}{3} + \frac{1}{1.2}\frac{x^5}{5} - \frac{1}{1.2.3}\frac{x^7}{7} \pm \frac{x^{2n-1}}{(2n-1)\lfloor n-1}\right]_0^{x_1}.$$

If x_1 be large, integrate $\int_{x_1}^\infty e^{-x^2}dx$, and subtract the result from $\sqrt{\pi}/2$.

Write $x^2 = y$, $2x\,dx = dy$, $dx = \tfrac{1}{2}y^{-\frac{1}{2}}dy$,
$$d\frac{y^m e^{ay}}{a} = \frac{y^m a e^{ay}}{a}dy + \frac{my^{m-1}e^{ay}}{a}dy\,;$$

where in the given case $m = \tfrac{1}{2}$, $a = -1$,
$$\int y^m e^{ay}dy = \frac{y^m e^{ay}}{a} - \frac{m}{a}\int y^{m-1}e^{ay}dy$$

$$-\frac{m}{a}\int y^{m-1}e^{ay}dy = -\frac{m}{a}\frac{y^{m-1}e^{ay}}{a} + \frac{m}{a}\frac{m-1}{a}\int y^{m-2}e^{ay}dy\,;$$

THE LAW OF ERROR.

substituting their values for y, m, and a,

$$\int_{x_1}^{\infty} e^{-x^2}dx = \left[\frac{e^{-x^2}}{\sqrt{2x}}\left\{1 - \frac{1}{2x^2} + \frac{1\cdot 3}{(2x^2)^2} - \frac{1\cdot 3\cdot 5}{(2x^2)^3} \pm \frac{1\cdot 3\cdot 5\cdots\overline{2n-3}}{(2x^2)^{n-1}}\right\}\right]_{x_1}^{\infty}.$$

The values thus found for these integrals are doubled and divided by $\sqrt{\pi}$ to give the ratio of the area of the curve between a positive and negative ordinate to the total area of the curve.

De Morgan, "Probability," *Enc. Metropol.*, gives very complete tables.

The integral found above expresses the relative probability of an error *less than* x, in very exact work the probability of an error $\pm x$ must be added, and the formula becomes

$$\frac{2}{\sqrt{\pi}}\left[\int_0^{x_1} e^{-x^2}dx + \frac{a}{2}e^{-x^2}\right],$$

where a is an infinitesimal thickness arbitrarily given to the ordinate, it is usual to add half the tabular difference to the entry, cf. "Probability," *Enc. Brit.*

The last column of Table I. is obtained by writing $0.476948x$ for x in the series.

6. If it could be assumed that the mean of the results gives the true value of the quantity observed, the sum of the squares of the residuals divided by their number would give the $(\text{E.M.S.})^2$.

But suppose, as is more probably the case, that the mean differs from the true value by a small quantity k, each residual v differs from the error $(v \pm k)$, and the sum of the squares of the true errors is

$$\sum[\overline{v\pm k}^2] \text{ or } \sum[v^2] \pm 2k\sum[v] + nk^2.$$

Since $\sum[v] = 0$ the second term vanishes, and the squares of the residuals must be corrected by a small quantity nk^2.

It is generally assumed that the best approximation is to suppose

$$nk^2 = (\text{E.M.S.})^2,$$

hence
$$n(\text{E.M.S.})^2 = \sum[v^2] + (\text{E.M.S.})^2;$$

$$\therefore (\text{E.M.S.})^2 = \frac{\sum[v^2]}{n-1}.$$

CHAPTER XVI.

THE WEIGHTS OF OBSERVATIONS AND THE GENERAL MEAN.

IT has been assumed hitherto that every observation is equally good, or that there is no reason for considering one better than another. In practice this is by no means always the case; some results may have been obtained under unfavourable conditions, by less skilful observers, or with less perfect instruments.

It is generally best to reject entirely observations, which there is any *real* reason for considering less satisfactory, but the rejection of doubtful results is a most delicate and difficult subject. Great care should be exercised, that it never takes place consciously or unconsciously in favour of a preconceived theory, by which the judgment is very liable to be biassed.

In some cases it may be better to retain the doubtful results, but to diminish their effect upon the final result.

THE GENERAL MEAN. 85

The opinion of the observer as to the value of different observations is expressed by multiplying each result by a number supposed to represent its relative weight; this is equivalent to assuming each observation to be repeated a number of times in proportion to its supposed accuracy. The fictitious results thus arrived at are dealt with just as though they were real, and the "general mean" is obtained by multiplying each observation by its weight, and dividing the sum by the sum of the weights. Theoretically, the weights of different series of observations of the same quantity are inversely proportional to the squares of their errors of mean square, but generally numbers supposed to represent the weights are arbitrarily assigned.

The following results were obtained by different observers, and methods for the "available oxygen" in a solution of potassium permanganate.

A 23.10; B 22.77, 22.55; C 22.58, 22.69;

D 22.27, 22.67, 22.52; E 22.71, 22.67.

If all the results are taken as of equal value, the mean is 22.653; but the values are not equal.

D were obtained by the hydrogen oxalate method, and are liable to error due to moisture, hence we may give them the arbitrary weight 2; B and C were obtained by the iron method with ordinary burettes, but C are rather closer together than B, hence we may assign the weight 4 to B, and 5 to C. E were obtained by the same observer as B with a weighing burette weight 6, and *A* was obtained by the teacher, the others being pupils, weight 8. Taking values above 22:

1.1	weight 8	8.80
$\left.\begin{array}{l}.77\\.55\end{array}\right\}$	,, 4	$\left\{\begin{array}{l}3.08\\2.20\end{array}\right.$
$\left.\begin{array}{l}.58\\.69\end{array}\right\}$	,, 5	$\left\{\begin{array}{l}2.90\\3.45\end{array}\right.$
$\left.\begin{array}{l}.27\\.67\\.52\end{array}\right\}$	,, 2	$\left\{\begin{array}{l}0.54\\1.34\\1.04\end{array}\right.$
$\left.\begin{array}{l}.71\\.67\end{array}\right\}$	,, 6	$\left\{\begin{array}{l}4.26\\4.02\end{array}\right.$
6.53	44	31.63

$\dfrac{31.63}{44} = 0.719$, hence on the above assumptions the general mean is 22.72.

In any large series of observations made with all possible care, some results, generally in excess of the number indicated by the law of error, show a comparatively large divergence from the mean. The general tendency is to reject such results as mistakes, or to weight them so that they produce little effect upon the final result. Both these methods are unsatisfactory, and a "criterion" has been proposed by Prof. Peirce:[1]

"Observations should be rejected when the probability of the system of errors obtained by retaining them is less than that of the system of errors obtained

THE GENERAL MEAN.

by their rejection, multiplied by the probability of making so many and no more abnormal observations."

This theory is not universally accepted, and the general application of it requires special tables, and presents some difficulties. A simple criterion for the rejection of one doubtful observation has been proposed by Prof. Chauvenet.

If there be n observations the error of mean square of which is e, assume that $T = \dfrac{2n-1}{2n}$, and obtain t from Table I. One observation with the residual a must be rejected or retained according as

$$a > \text{ or } < 0.6745 te.$$

Thus if the E.M.S. of fifteen observations be $0''.572$, $T = (30-1)/30 = 0.9667$ and $t = 3.155$.

One observation with a residual greater than

$$0.6745 \times 3.155 \times 0''.572 = 1''.217$$

must be rejected.

REFERENCES AND NOTES.

1. Chauvenet: *Method of Least Squares*, p. 558, gives a full account of the method with the necessary tables.
Peirce: *Astronomical Journal* (Camb. Mass.), ii., 161.
Airy, iv. 137, doubts the validity.
Winloch, iv. 145 answers Airy.
Glaisher, *loc. cit.* agrees with Airy.

CHAPTER XVII.

THE METHOD OF LEAST SQUARES.

THE more general problem, to determine the most probable values of a number of unknown quantities, of which the observational results are functions, is so complicated as to be practically insoluble, unless the observational equations are of, or can by some artifice be reduced to, the linear form. The most general method of effecting this is to assume approximate values for the unknowns, and to solve for small corrections to be applied to the assumed values.

If the last terms of the equations have not the same probable error, each equation must be multiplied by the measure of precision or square root of the weight of the observational value, which renders the probable error of each the same. If the number of the equations is equal to that of the unknowns only one value of each of the latter can be obtained. If there are more equations than unknowns, the equations must be so combined that the probable

THE METHOD OF LEAST SQUARES. 89

errors of the deduced values of the unknowns shall be minima, or that the sum of the squares of the errors remaining after correction for the deduced values of the unknowns shall be minimum.

Both these results are reached as follows:

Multiply each equation by the coefficient of x in it, and add all the equations to obtain the "normal equation for x." Proceed in the same way to find normal equations for y, z and the other unknowns. The normal equations are equal in number to the unknowns, and when solved give the most probable values for them.

Thus to find the most probable values for x, y, z from the five equations

$$x=10, \quad y-x=7, \quad x-z=2, \quad y=18, \quad y-z=9.$$

$$\begin{array}{lll} x= 10 & y=18 & -y+z=-9 \\ -y+x=-7 & y-x=7 & -x+z=-2 \\ -z+x=2 & y-z=9 & \\ \hline 3x-y-z=5 & -x+3y-z=34 & -x-y+2z=-11 \end{array}$$

are the normal equations for x, y, and z.

Solving by any method we find

$$x=10\tfrac{3}{8}, \quad y=17\tfrac{5}{8}, \quad z=8\tfrac{1}{2},$$

as the most probable values for the unknowns.

CHAPTER XVIII.

CONDITIONED EQUATIONS.

THE unknown quantities are sometimes subject to conditions, which, if exact, they must rigorously satisfy. Thus all the angles measured in one plane round a point must be equal to 360°, all the angles of a plane triangle must together equal 180°. The sum of the weights of the constituents of a body must be equal to the weight of the body.

In such a case to obtain the most probable values one of the unknowns is eliminated by the aid of the conditional equation; the normal equations of the other unknowns are then formed and solved in the ordinary way.

Suppose that the angles between the normals to the faces of a triangular prismatic crystal have been found to be $\theta = 121°\ 10'$, $\phi = 119°\ 36'$, $\psi = 119°\ 30'$, the sum of which is 360° 16', instead of the true value 360°.

Write $\theta + \phi = 360 - \psi = 240°\ 30'$, and there are now three equations to determine θ and ϕ, the normal equations are

$$2\theta + \phi = 361° \ 40', \quad \theta + 2\phi = 360° \ 6',$$
$$\therefore \theta = 121° \ 4' \ 40'', \quad \phi = 119° \ 30' \ 40'', \quad \psi = 119° \ 24' \ 40''.$$

In all cases it is safer and more satisfactory to give both the uncorrected observational results and those which have been reduced for further use; since the former furnish the real test of the accuracy and fidelity of the experiments, and obviate any unconscious tendency to "cook."

The percentage composition of a substance must theoretically add up to 100, some chemists multiply or divide their results to get rid of discrepancy, or leaving one result undetermined, calculate it by difference. In accurate work both these practices should be avoided. Far too often a valuable product or a considerable error has been overlooked by being reckoned as a difference. Again, in stating the results of an analysis, it is well to give the absolute amount, or at least the percentage of each substance found, and not, as is too often done, only the percentages of the substances, which the radicals actually found are supposed to form. In a water analysis the amount of sodium, magnesium, and chlorine present should be stated as well as the percentage of sodium chloride and of magnesium chloride.

CHAPTER XIX.

GENERAL FORMULAE.

IT frequently happens that a series of values of two or more inter-dependent quantities have been determined by observation, and it is required to find a general expression for one in terms of the other. A general formula decreases the burden on the memory, and enables us to find values intermediate between those given by observation, or even to a certain extent beyond them.

The problem is essentially an indeterminate one, since various formulae can be found, which will express the results within the limits of the errors of observation; but the formulae may be so cumbersome as to be practically useless.

In general it is very difficult to find a suitable formula if more than two variable quantities are considered at once; though two or more known formulae may often be used simultaneously.

Thus, if we wish to find the change in volume of a mass of gas produced by change of temperature and

GENERAL FORMULAE.   93

change of pressure, keeping the temperature constant we find the relation between volume and pressure, and then keeping the pressure constant we find the relation between volume and temperature. Had all three quantities been allowed to vary together, it would have been difficult to determine the two formulae at once, though, when found, they are easily used in conjunction.

There are three common methods of attempting to find a general formula, which are usually tried in the following order: by deduction from general principles the form of the function connecting the quantities is determined, the constants in the formula are then found from the experimental results; by graphical methods, which essentially consist in drawing a curve through the observational results suitably plotted out; by the trial of a succession of purely empirical formulae, until one is found which expresses the results within the limits of error and with sufficient convenience.

If no sufficiently convenient formula can be found, and the results are frequently required, they are generally tabulated.

Considerable practice and skill is required in the application of each of these methods, and no rules can be given which will apply in every case, or even afford much assistance. All the observational results must be *fairly* represented by the formula, but it must not be expected that they can be *exactly* represented. It is usually assumed that the variable and variant change continuously, unless a marked change of state or conditions occur, but this general rule must not be too much relied upon, as we may not know when a sufficient change takes place.

Suppose a body is dropped from a height down a coal-mine, the attraction of the earth varies at first inversely as the square of the distance of the particle from the centre, but, when the surface of the earth (considered homogeneous) is passed, the attraction varies directly as the distance of the particle from the centre.

If a ray of light be passed from water to air, and the angle between the ray and the normal be gradually increased, at 48° 30′ the light ceases to emerge at all.

CHAPTER XX.

THE DEDUCTIVE METHOD.

Too little is known as to the ultimate constitution of matter to enable us safely to deduce formulae in the majority of cases which occur in physics and chemistry. Even in mechanics and astronomy the method must be used with great caution, and the results obtained are in general only approximate.

It has been applied to determine the equations representing planetary orbits, the figure of the earth, atmospheric refraction, and the measurement of heights by the barometer.

Suppose it be desired to find a formula giving the value of the attraction of the earth (g) in different latitudes (ϕ). Assuming the earth to be a homogeneous spheroid of small eccentricity (e), we find

$$g_\phi = g_0 \left(1 + \frac{e^2}{10} \sin^2\phi\right).$$

But the earth is not a true spheroid, it increases rapidly in density towards the centre, and it is in rapid rotation; hence the above formula can be only a rough approximation to the truth, though we may adopt the form of it. In fact the value of $e^2/10$ is about 0.000665, while that of the coefficient found by observation is 0.00534.

The exact representation of any phenomenon or change, which actually occurs, by mathematical symbols and processes is impossible; an analogous but much simplified case must be assumed and worked out, various small corrections being introduced, when possible, to fit the results to the actual facts.

Thus it is frequently assumed that solids are rigid and without weight, that liquids are without internal friction, that gases are perfect, that the earth is at rest, and that gravity acts towards its geometrical centre. Material particles are treated merely as centres or points of application of force, and small surfaces of emission as though they were points.

It is a good and humbling exercise to compare the theoretical action of a lever of the first kind with the actual action of a crowbar in lifting a stone. How many small changes are neglected?

CHAPTER XXI.

GRAPHICAL METHODS.

OWING to their simplicity and convenience, graphical methods[1] are very frequently used for three different purposes : to record and connect a series of separate observations; by the aid of special instruments to keep a continuous record of a perpetually changing quantity; or, in cases where great accuracy is not required, to replace calculation by the measurement of lines which represent quantities and directions.

Graphical methods of the third kind have recently been introduced with considerable success into practical mechanics and engineering, but apparently have not yet been found of great utility in physics and chemistry, hence they may be omitted from further consideration. The other two applications of graphical methods must be treated rather more fully.

The observational results are plotted by pricks, dots, or cross lines on a sheet of metal or paper,[2] ruled into squares of a convenient size. A curve is drawn by the free hand or by the aid of a flexible lath of

98   *NOTES ON OBSERVATIONS.*

wood or steel, so as to pass as evenly as possible among the dots. The curve is assumed to give the most reliable expression for the general formula connecting the experimental results, and deviations from it are assumed to be due to errors of observation.

The problem of drawing a curve through any given points is essentially an indeterminate one, since any number of curves can be drawn to pass through all the points. Hence other assumptions, such as the following, are introduced. The curve with the fewest changes of curvature, which passes through many of the points, and within a comparatively small distance, representing the probable error of the observation, of the majority of the rest must be selected. Nearly an equal number of the experimental results should lie on each side of each small portion of the curve. Very few experimental points must be at a comparatively considerable distance from the curve selected.

It is evident that the form of the curve adopted depends much upon the judgment of the operator; and that two operators might represent the same series of results by different curves, especially if they considered the probable error of the observations, as having very different values. One might consider the probable error as comparatively large, and prefer a simple curve which did not represent the experiments very closely. Another might consider the probable error very small, and prefer a more complicated curve passing more nearly among the experimental points.

A third might consider it more in accordance with the experimental results to use instead of the complicated curve two or more simpler curves which intersected one another.

Imagine a steel wire, a square millimetre in section, and 20 metres long, to which a light scale pan is attached, suspended vertically. If a weight of 1 kilo. be placed in the pan, the wire stretches 1 MM. A second weight of 1 kilo. produces a second stretch of 1 MM., and so on. This connection, the simplest we can have between any two quantities, is expressed by Hooke's law: "Ut tensio sic vis."

Draw two straight lines at right angles, which meet in the point O, call the horizontal line (which is generally by convention drawn from left to right) the x-axis, and the vertical line the y-axis. Measure off units of weight parallel to x horizontally, and units of increase of length parallel to y vertically.

We thus obtain a series of intersections of the x and y ordinates, through which the straight line OP can be drawn, this line represents the connection between the stress and the strain, or the result of our experiments. If we wish to know the strain corresponding to any stress, or the stress corresponding to any strain, we have only to draw the ordinate representing the given quantity, and then to measure the ordinate representing the unknown quantity. In this special case the two ordinates are equal.

A very small proportion of the observational results are accurate, even so far as our means of measurement extend, hence, in the great majority of cases, the points, which represent them, do not lie upon, but upon one side or the other of the straight line. A test of the accuracy of the observations and of the care with which the line is drawn is afforded by the equal distribution of the points on the two sides of each part of the line, and by the relatively short distances they lie away from it.

Invaluable as the graphical method is in very many cases, and useful as it always is, in extremely accurate work the errors introduced by it become comparable with those due to the observations, and therefore it cannot be entirely trusted.

A graphical reduction comprises five operations, each liable to error. Measurement of the abscissae, measurement of the ordinates, drawing the curve, measurement of the abscissa, measurement of the ordinate of the new value required.

There is a certain limit of size beyond which increase of the scale does not conduce to increase in accuracy, and this limit seems to be about a square metre, so that the unaided eyesight can only read to about 1/2000.

No result is accurately represented by a point, but more truthfully by a circle, the radius of which is equal to the "probable error" of the observation expressed on the same scale as the diagram. So far as that one observation goes, no evidence is afforded against any curve which cuts the circle.

It is very difficult to assign the accuracy attainable

by this method. Probably most draftsmen would consider it greater than 1/1000 of the quantity measured, or, that the third figure is accurately represented, few would claim for their work an accuracy of 1/10000, or that the fourth figure is exactly true.

Suppose that an ordinate of length a makes a small angle, the circular measure of which is θ, with the true normal to the base, the error in the position of the extremity of the ordinate is nearly $a\theta$, or if θ be 1', about 1/3400 of the length of the ordinate.

When the quantity observed varies rapidly or irregularly, a continuous series of observations is required to record its changes, hence much trouble is saved, and frequently accuracy is gained by making the instrument self-recording.

Many different forms of recording instruments are in use, but almost all depend upon two actions.

Instead of reading the instrument, the result is photographed upon a sheet of sensitive paper, either stationary or slowly moving by clock-work. When developed the sheet gives a continuous record of the observations. An ink-pencil connected with the instrument moves over a sheet of paper, either stationary, connected with some other moving part of the instrument, or with independent clock-work; the paper may be flat or more usually rolled round a cylinder. The paper is sometimes replaced by smoked glass, and the ink-pencil by a pointer.

One of the simplest recording instruments is Jordan's Sunshine Recorder, in which sunshine passing through a slit in a brass box acts upon a sheet of sensitive paper placed below. In this case the sun itself acts as the clock, and simply records the times at which it is powerful enough to affect the sensitive paper.

In Watt's Indicator, a pencil connected with a piston pressed down by a spring, records the pressure of the steam in the cylinder of an engine on a sheet of paper rolled on a drum, which rotates in connection with the piston rod of the engine. The closed curve traced out records at any time the pressure and the volume of the steam in the cylinder of the engine. The area of the curve measured by a planimeter gives the work done by the steam.

The marvellous instruments invented by Lord Kelvin[3] to record and predict the height of the tide at any given place, furnish a more complicated but an almost theoretically perfect example.

The tide raises or lowers a float which records its position by means of an ink-pencil upon a sheet of paper rolled on a cylinder rotated by clock-work.

When the sheet is unrolled the tide-gauge thus gives a continuous record of the state of the tide, as a sinuous line traced on a flat sheet of squared paper.

This sheet of paper is then put into a very complicated machine, the harmonic analyser, by which the observational results are, as it were, broken up, and referred back to the action of the moon, the sun, and eight minor causes.

On supplying the ten data obtained from the analyser to a third machine, the tide-predictor, and turning a handle a sinuous curve is produced similar in character to that given by the tide-gauge, but the new curve shows the state of the tide at any future instead of at any past time!

REFERENCES AND NOTES.

1. Hele Shaw: *B.A. Reports*, 1892, gives an excellent account of graphic methods.
2. A very convenient paper of French manufacture consists of sheets a metre square ruled into millimetre squares by dotted lines. Since the dots are 0.2 mm. apart, 0.1 mm. can be estimated with fair accuracy.
3. Kelvin: *Popular Lectures III.*, "The Tides."

CHAPTER XXII.

EMPIRICAL FORMULAE.

THERE are no general rules for finding empirical formulae; success depends upon tact and experience. If the value of the function increases or decreases continuously as x increases, some algebraic formula will probably be suitable, but if the value of the function increases and decreases alternately, a series of sines or cosines is suggested. When the form of the function has been chosen by the aid of theory, analogy, or the shape of the plotted curve, the constants are determined from the results of observation, and the calculated are compared with the observed values.

In many cases the value of the variant may be expressed by one or more terms of the series

$$u_x = \phi(x) = a + bx + cx^2 + dx^3 + \ldots,$$

where a, b, c, d are constants which may be positive, zero, or negative.

The assumption that $\phi(x)$ can be expressed in a

series of ascending powers of x is by no means always true, as it only applies to some mathematical functions.

It is often found in practice that the number of terms required renders the series unmanageable. There are several methods of determining the constants, one of the simplest of which is as follows.

By subtracting the constant term in the series, reduce each equation to the form

$$u_1 = bx_1 + cx_1^2 + dx_1^3.$$

Subtract each equation from the one below it, divide by $\overline{x_2 - x_1}$ and similar terms, write

$$\delta u_1 = \frac{u_2 - u_1}{x_2 - x_1} = b + \overline{x_2 + x_1}\ c + \overline{x_2^2 + x_2 x_1 + x_1^2}\ d,$$

$$\delta u_2 = \frac{u_3 - u_2}{x_3 - x_2} = b + \overline{x_3 + x_2}\ c + \overline{x_3^2 + x_3 x_2 + x_2^2}\ d,$$

$$\delta u_3 = \frac{u_4 - u_3}{x_4 - x_3} = b + \overline{x_4 + x_3}\ c + \overline{x_4^2 + x_4 x_3 + x_3^2}\ d;$$

subtract again, and divide by $\overline{x_3 - x_1}$ and similar terms,

$$\delta^2 u_1 = \frac{\delta u_2 - \delta u_1}{x_3 - x_1} = c + \overline{x_3 + x_2 + x_1}\ d,$$

$$\delta^2 u_2 = \frac{\delta u_3 - \delta u_2}{x_4 - x_2} = c + \overline{x_4 + x_3 + x_2}\ d;$$

subtracting again, and dividing by $\overline{x_4 - x_1}$,

$$\delta^3 u_1 = d.$$

Substituting successively in the equations

$$d = \delta^3 u_1,$$
$$c = \delta^2 u_1 - \overline{x_3 + x_2 + x_1}\, \delta^3 u_1,$$
$$b = \delta u_1 - \overline{x_2 + x_1}\, \delta^2 u_1 + \overline{x_3 x_2 + x_3 x_1 + x_2 x_1}\, \delta^3 u_1,$$
$$a = u - bx - cx^2 - dx^3.$$

When, as occasionally happens, x_1, x_2, x_3 are in arithmetical progression with a common difference h:

$$\delta u_1 = \frac{u_2 - u_1}{h},$$

$$\delta^2 u_1 = \frac{\delta u_2 - \delta u_1}{2h^2},$$

$$\delta^3 u_1 = \frac{\delta^2 u_3 - \delta^2 u_1}{6h^3}.$$

Owing to errors of observation the differences found are never exact. Dr. John Hopkinson[1] has proposed the following convenient method for obtaining probable results: Add together the various values of each quantity obtained, and take the mean as the most probable value.

He makes the following remarks: "Though the method has no theoretical basis to rest upon, it is comparatively easy and should give the same results in all hands, while any method of plotting and curve drawing introduces a further series of observations liable to personal error.

"There are three advantages:

"(a) The sum of the differences of the observed and calculated values = 0.

EMPIRICAL FORMULAE.

"(*b*) If the values in the last column of differences regularly increase or decrease, another term must be added to the equation.

"(*c*) If the differences are very irregular, there is a want of accuracy in the observations, or the theoretical equation is carried further than the experiments warrant."

If N be the number of the observational equations, each differencing removes one, so that going as far as $\delta^3 u$ there are really only $N-3=n$ complete equations to deal with. The following separate values must be found from each equation, added together and divided by n, so as to obtain

$$\Sigma x/n, \quad \Sigma x^2/n, \quad \Sigma x^3/n,$$
$$\Sigma u/n, \quad \Sigma \delta u/n, \quad \Sigma \delta^2 u/n, \quad \Sigma \delta^3 u/n;$$

also the following rather complicated sums which may be a little simplified:

$$\Sigma(\overline{x_1+x_2}+\overline{x_2+x_3}+\ldots+\overline{x_n+x_{n+1}})/n$$
$$=\frac{2\overline{\Sigma x_1 \text{ to } x_n}+x_{n+1}-x_1}{n},$$

$$\Sigma(\overline{x_1+x_2+x_3}+\ldots+\overline{x_n+x_{n+1}+x_{n+2}})/n$$
$$=\frac{3\overline{\Sigma x_1 \text{ to } x_n}+x_{n+2}+2x_{n+1}-x_2-2x_1}{n},$$

$$\Sigma(\overline{x_1 x_2+x_2 x_3+x_3 x_1}+\ldots+\overline{x_n x_{n+1}+x_{n+1}x_{n+2}+x_{n+2}x_n})/n$$
$$=\frac{\overline{\Sigma x_1 x_3 \text{ to } x_n x_{n+2}}+2\overline{\Sigma x_1 x_2 \text{ to } x_n x_{n+1}}+x_{n+1}x_{n+2}-x_1 x_2}{n}.$$

As an example of the reduction of a series of

experimental results, we may examine the apparent expansion of phosphorous oxide in a glass dilatometer determined by Dr. Thorpe and Mr. Tutton.[2]

$t°$ C.	$t - 27.1$	u obs.	u calc.	Diff.	
27.10	0.00	1.0000	1.0000	0.0000	
36.07	8.97	1.0078	1.0078		0
47.66	20.56	1.0184	1.0180	−	4
63.20	36.10	1.0317	1.0316	−	1
77.06	49.96	1.0443	1.0441	−	2
89.48	62.38	1.0563	1.0557	−	6
103.99	76.89	1.0693	1.0693		0
117.68	90.58	1.0826	1.0827	+	1
130.70	103.60	1.0962	1.0959	−	3
140.30	113.20	1.1063	1.1058	−	5

From the observed results the authors find the general formula:

$$u = 1 + 0.00088824(t - 27.1) - 0.000000013873(t - 27.1)^2 + 0.0000000038446(t - 27.1)^3,$$

from which the fourth and fifth columns in the table are calculated.

We may select the three widely separated equations marked by an asterisk to calculate the constants b, c, d in the assumed expansion. Write x for $t - 27.1$. Subtracting 1 and dividing through by the coefficient of b.

EMPIRICAL FORMULAE.   109

$$b + 113.2c + 12814.24d = \frac{.1063}{113.2} = .000939046,$$

$$b + 62.38c + 3891.2644d = \frac{.0563}{62.38} = .000902533,$$

$$b + 8.97c + 80.4609d = \frac{.0078}{8.97} = .000869565.$$

Hence

$$c + 175.58d = \frac{.000036513}{50.82} = 0.000000718477$$

$$c + 71.35d = \frac{.000032968}{53.41} = 0.000000617263$$

$$104.23d = 0.000000101214.$$
$$d = 0.000000000971,$$
$$c = 0.000000548,$$
$$b = 0.000864581,$$
$$a = 1,$$

from which the following table is calculated:

x	$1 + bx$	cx^2	dx^3	u calc.	Diff.
0.	1.000000	.000000	.000000	1.00000	0.00000
*8.97	1.007755	.000044	.0000007	1.00780	0
20.56	1.017780	.000330	.000008	1.01812	− 28
36.10	1.031211	.000714	.000046	1.03197	27
49.96	1.043194	.001368	.000121	1.04483	53
*62.38	1.053933	.002132	.000236	1.05630	0
76.89	1.066556	.003247	.000443	1.07025	95
90.58	1.078314	.004496	.000722	1.08353	93
103.60	1.089571	.005882	.001080	1.09653	33
*113.20	1.097871	.007022	.001408	1.10630	0

Though the calculated agree fairly well with the observed results, it is probable that the experimental results are low in the two last selected equations, or that c and d ought to be slightly larger. Further, d is so small, that probably the term involving it might be neglected. In fact

$$u = 1 + 0.000859x + 0.00000067x^2,$$

though it makes the individual differences rather larger, renders them alternately positive and negative with a smaller sum.

As another example, we may apply Dr. Hopkinson's method to find the coefficients in the equation

$$u = a + bx + cx^2$$

from the same experiments.

Since the values of $\delta^2 u$ are irregular and two of them are negative, the experimental values are not sufficiently accurate to require a fourth term.

From the tabular values:

$$c = \overline{\delta^2 u} = 0.00000089336$$

$$b = \overline{\delta u} - \overline{\delta^2 u \Sigma x_1 + x_2}/n = \overline{\delta u} - \overline{\delta^2 u}\frac{2 \times 345.44 + 103.6}{8}$$

$$= 0.0009284 1 - 0.00000089336 \times 99.31$$

$$= 0.00083969,$$

$$\bar{x} = 43.18, \quad \overline{x^2} = 2788.8074,$$

$$a = \bar{u} - b\bar{x} - c\overline{x^2}$$

$$= 1.0388 - 0.00083969 \times 43.18$$
$$\qquad - 0.00000089336 \times 2788.8074$$

$$= 1.0388 - 0.038749 = 1.000051.$$

EMPIRICAL FORMULAE.

x	x^3	u obs.	δu	$\delta^2 u$	u calc.	Diff.
0.00	0.0000	1.0000	0086956	0.21896	1.00005	+.00005
8.97	80.4609	1.0078	91458	− 21644	1.00757	− 23
20.56	422.7136	1.0184	85586	18105	1.01769	− 71
36.10	1303.2100	1.0317	90909	21723	1.03153	− 17
49.96	2496.0016	1.0443	96618	− 26086	1.04423	− 7
62.38	3891.2644	1.0563	89593	26801	1.05591	− 39
76.89	5912.0721	1.0693	97151	27341	1.06989	+ 59
90.58	8204.7364	1.0826	00104454	03333	1.08344	+ 84
103.60		1.0962	00105208	0.71469	1.09662	+ 42
113.20		1.1063		0.89336	1.10654	+.00024
345.44	22310.4590	8.3104	00742725			+.00057
43.18	2788.8074	1.0388	00092841			

Hence we find the values given under u *calc.* from the equation

$$u = 1.00005 + 0.0008396 9x + 0.00000089336x^2.$$

Since the sum of the differences is $+0.00057$, 0.00005 may be omitted in the first term, and probably the last figures of the other coefficients. In fact

$$u = 1 + 0.00084x + 0.00000089x^2.$$

gives a very approximate result, the sum of the differences being only 0.00002.

It may be noted that

$$\Sigma x^2 / n \div (\Sigma x/n)^2 = \frac{4585.7}{3160.6} = 1.45$$

instead of $\pi/2$. This result, and the two negative second differences, probably point to some accidental error in two of the experiments.

The two methods just given are usually the most simple, and suffice to determine the constants with considerable accuracy. But when three equations as widely separated as the experimental results permit have been selected and reduced to the form

$$bx_1 + cx_1^2 + dx_1^3 - u_1 = 0,$$
$$bx_2 + cx_2^2 + dx_2^3 - u_2 = 0,$$
$$bx_3 + cx_3^2 + dx_3^3 - u_3 = 0,$$

the constants may be determined by the theory of

EMPIRICAL FORMULAE.

determinants, by the rule of cross-multiplication, or by indeterminate multipliers, which gives:

$$b = \frac{u_1(x_2^2 x_3^3 - x_2^3 x_3^2) + u_2(x_1^3 x_3^2 - x_1^2 x_3^3) + u_3(x_1^2 x_2^3 - x_1^3 x_2^2)}{x_1(x_2^2 x_3^3 - x_2^3 x_3^2) + x_2(x_1^3 x_3^2 - x_1^2 x_3^3) + x_3(x_1^2 x_2^3 - x_1^3 x_2^2)},$$

$$c = \frac{u_1(x_2^3 x_3 - x_2 x_3^3) + u_2(x_1 x_3^3 - x_1^3 x_3) + u_3(x_1^3 x_2 - x_1 x_2^3)}{x_1^2(x_2^3 x_3 - x_2 x_3^3) + x_2^2(x_1 x_3^3 - x_1^3 x_3) + x_3^2(x_1^3 x_2 - x_1 x_2^3)},$$

$$d = \frac{u_1(x_2 x_3^2 - x_2^2 x_3) + u_2(x_1^2 x_3 - x_1 x_3^2) + u_3(x_1 x_2^2 - x_1^2 x_2)}{x_1^3(x_2 x_3^2 - x_2^2 x_3) + x_2^3(x_1^2 x_3 - x_1 x_3^2) + x_3^3(x_1 x_2^2 - x_1^2 x_2)}.$$

Were all the experimental results absolutely accurate, the same values for the constants would be found whatever experimental results were chosen, but owing to errors of observation the values found are never identical.

The most probable values are found as follows:

Reduce all the experimental equations to the form

$$bx_1 + cx_1^2 + dx_1^3 - u_1 = 0.$$

If the observational results (u_1 etc.) have not the same probable error, multiply each equation by the reciprocal of the probable error, by the measure of precision, or by the square root of the weight of the observed value in it. Equations thus prepared are spoken of as equations of condition.

Multiply each equation of condition by the coefficient of b in it, and add all the equations together; the resulting "normal equation for b" gives, when solved, the most probable value for b. Proceed in the same way by multiplying each equation by their coefficients to find normal equations for c and d.

Writing
$$X^2 \text{ for } \Sigma x_1^2 + x_2^2 + \text{etc.,}$$
$$X^3 \text{ for } \Sigma x_1^3 + x_2^3 + \text{etc.,}$$
$$\dots\dots\dots\dots\dots\dots\dots\dots\dots\dots\dots$$
$$U^1 \text{ for } \Sigma u_1 x_1 + u_2 x_2 + \text{etc.,}$$
$$U^2 \text{ for } \Sigma u_1 x_1^2 + u_2 x_2^2 + \text{etc. ;}$$
the normal equations are
$$X^2 b + X^3 c + X^4 d - U^1 = 0,$$
$$X^3 b + X^4 c + X^5 d - U^2 = 0,$$
$$X^4 b + X^5 c + X^6 d - U^3 = 0.$$

After carrying the arithmetical reduction as far as possible, the necessarily tedious solution may be effected by substituting X^2 for x_1 etc., in the formula given above. In physics and chemistry it is rarely necessary to employ the symmetrical but still tedious method of solution due to Gauss. The simple example which has been used before will serve to exemplify the method. The following values are required, X^2, X^3, X^4, U^1, U^2.

$$45857.659 b + 4184166.79 c - 42.127727 = 0,$$
$$4184166.79 b + 404927806 c - 3861.45076 = 0,$$
$$b + 91.2448845 c - 0.00091866283 = 0,$$
$$b + 96.7762105 c - 0.00092287208 = 0,$$
$$c = \frac{.00000420925}{5.531326} = 0.000000760985,$$
$$b = 0.00091866283 - 0.00006943599$$
$$= 0.00084922684 ;$$
$$\therefore u = 1 + 0.00084922684 x + 0.000000760985 x^2.$$

EMPIRICAL FORMULAE.

x	x^2	x^3	x^4	$u-1$	$(u-1)x$	$(u-1)x^2$
0.00	0.0000	0.000000	0.00	.0000	0.000000	0.000000
8.97	80.4609	721.734273	6473.81	.0078	0.069966	0.627684
20.56	422.7136	8690.991616	178675.29	.0184	0.378304	7.777930
36.10	1303.2100	47045.881000	1698356.30	.0317	1.144370	41.311757
49.96	2496.0016	124700.239936	6230016.	.0443	2.213228	110.572871
62.38	3891.2644	242737.073272	15141827.	.0563	3.511994	219.078186
76.89	5912.0721	454579.223769	34952572.	.0693	5.328477	409.706596
90.58	8204.7364	743185.023112	67317594.	.0826	7.481908	677.711228
103.60	10732.9600	1111934.656000	115197289.	.0962	9.966320	1032.510752
113.20	12814.2400	1450571.968000	164205003.	.1063	12.033160	1362.153712
	45857.6590	4184166.790978	404927806.		42.127727	3861.450716

Since $x<114$, and the experimental results are only true to 0.0001, $0.00005/114=0.0000004$ may be neglected or b taken as 0.000849; also $x^2<13000$, and $0.00005/13000=0.000000004$ may be neglected or c taken as 0.00000076.

We have then found the following equations:

$$u = 1 + 0.00088824x - 0.00000013873x^2 + 0.0000000038446x^3,$$

$$u = 1 + 0.00086458Ix + 0.000000548x^2 + 0.0000000009711x^3,$$

$$u = 1 + 0.000859x + 0.00000067x^2,$$

$$u = 1 + 0.00084x + 0.00000089x^2,$$

$$u = 1 + 0.000849x + 0.00000076x^2.$$

It is evident from the above that several different formulae can be found, which more or less adequately represent any given series of observational values. It is frequently a matter of considerable difficulty to decide which formula to select as most convenient and sufficiently accurate. Excluding the first, the last formula in the above case is to be preferred for theoretical reasons.

To complete the reduction in the above case it is necessary to allow for the expansion of the glass dilatometer, which must be determined by another series of observations.

If the equations representing the apparent expansion of the liquid and that of the envelope are identical in form, and in each case the constant term is unity and the coefficients small and decreasing;

when great accuracy is not required, it is sufficient to add the coefficients and obtain an equation of the form
$$u = 1 + (b_1 + b_2)x + (c_1 + c_2)x^2 + (d_1 + d_2)x^3,$$
otherwise the two results must be obtained at each temperature and a new general formula must be calculated from the sums of the two values.

From the general formula a table of the values of the variant is calculated for equal increments of the variable. It is convenient to take these increments so small that intermediate values of the variant can be found by simple proportion without the trouble of applying an interpolation or the general formula.

If the formula involving only constants and ascending whole powers of the variable fails, and theory affords no clue, even the greatest tact and experience may not succeed in finding any satisfactory expression.

Many formulae have been suggested to express the relation between the temperature and the tension of aqueous vapour; the one generally preferred, which is due to Biot, is complicated and tedious to use. Several formulae have been employed by different experimentalists to express the alteration in the resistance of platinum with change of temperature.

The assumption of a new variable or variant connected with the one observed may give a less complicated function; thus the reciprocal or logarithm of the variable may sometimes be more easily formulated than the variable itself.

The observational values may differ from those given by a simple equation only for certain values, which may be corrected by the addition of a term, which becomes less than the errors of observation above and below the required values. Thus a term of the form $\dfrac{n}{e^x + e^{-x}}$ or $\dfrac{n}{2} \operatorname{sech} x$ becomes insensible for positive and negative values of x which are large compared to n. If $n = 2$, the function becomes 1 for $x = 0$ and < 0.0001 when $x = \pm 10$.

Occasionally when no formula can be found to represent the connection between two quantities over a long range, each part of the range may be more easily represented by a separate equation. Thus Kopp gave four equations to represent the expansion of water from 0° to 100° C., each of which only held good for 25°.

If after equal increments in the value of x the value of the function $y = \phi(x)$ is found to recur, the form of the function is probably trigonometrical. The simplest form of such an equation is

$$y = \sin x$$

when $x = 0$, $\dfrac{\pi}{4}$, $\dfrac{2\pi}{4}$, $\dfrac{3\pi}{4}$, π, $\dfrac{5\pi}{4}$, $\dfrac{6\pi}{4}$, $\dfrac{7\pi}{4}$, 2π;

$y = 0$, 0.7, 1, 0.7, 0, -0.7, -1, -0.7, 0.

The curve is symmetrical and sinuous, the loops are alternately above and below the x-axis, cutting it at angles $\pi/4$. The length of the base of each

EMPIRICAL FORMULAE.

loop is $\pi = 3.14$, and its height, 1; the length of the curved loop is 3.8212, and its area, 2.

Comparing the curve with a cycloid traced out by a point on the circumference of a circle of diameter 1 which rolls through an angle θ along the base,

$$x = \tfrac{1}{2}(\theta - \sin \theta),$$
$$y = \tfrac{1}{2}(1 - \cos \theta).$$

The length of the cycloid is 4 and its area 2.356; it meets the base at right angles, and lies outside the sine curve, meeting it in the points

$$(0, 0), \quad \left(\frac{\pi}{2}, 1\right), \quad (\pi, 0).$$

The sine curve expresses the passage of a radiant vibration through the ether. A constant which multiplies the sine is spoken of as the amplitude of the vibration, and the recurrent value as the phase of the vibration.

It rarely happens in practice that the loop of the sine curve is symmetrical, and can therefore be expressed by a single term. The more general problem is to find a curve which will pass through n points, the abscissae of which are unequal. This can generally be effected by taking n terms of the series

$$y = A \sin x + B \sin 2x + C \sin 3x + \text{etc.},$$

and determining the constants.

The simplest way to draw a curve representing a sine series is to plot each term successively and add the new ordinate to the sum of the ordinates found from the earlier terms.

To take an extreme case, since

$$1 = \frac{4}{\pi}\left(\sin x + \frac{1}{3}\sin 3x + \frac{1}{5}\sin 5x + \text{to } \infty\right),$$

$$y = \sin x + \frac{1}{3}\sin 3x + \frac{1}{5}\sin 5x + \text{etc.},$$

gradually approximates to a rectangle, the base of which is the x-axis, the opposite side of which is at a distance $y = \pi/4$ from it.

The cosine curve $y = \cos x$ is similar to the sine curve, except that the phase differs by $\pi/2$, or the origin has been moved along the x-axis. The complete expression

$$\cos x + \cos 2x + \cos 3x + \ldots \cos nx = -\frac{1}{2} + \frac{1}{2}\frac{\sin(2n+1)\frac{x}{2}}{\sin\frac{x}{2}}.$$

The curve $y = \text{vers. } x = 1 - \cos x$ is also similar to the sine curve except that it lies above the x-axis, the origin having been moved to $\left(-\frac{\pi}{2}, -1\right)$.

Any discussion of Fourier's series is quite beyond the scope of elementary notes, and intimately connected with spherical harmonics.[3]

REFERENCES AND NOTES.

1. Hopkinson: *Messenger of Mathematics*, 1872, ii., 65.
2. Thorpe and Tutton: *Journal of the Chemical Society*, June, 1890, p. 559.

Owing to a misprint or a slip in arithmetic my calculation does not agree so closely with the observed values as that given by the authors.

This example, owing to special difficulties, must not be considered typical of the highest attainable accuracy, but merely as serving to briefly exemplify some methods of reduction. For examples of greater accuracy and length, consult—

Travaux et Mémoires. Bureau International des Poids et Mesures.

Thorpe and Rücker: "On the Expansion of Sea-Water by Heat." *Phil. Trans.*, clxvi. pt. 2, p. 405.

3. Byerly: *Fourier's Series and Spherical Harmonics*, Boston. 1893.
Gray and Matthews: *A Treatise on Bessel's Functions*.

TABLE I.

x or t	$\frac{2}{\sqrt{\pi}}\int_0^x e^{-x^2}dx$	$\frac{2}{\sqrt{\pi}}\int_0^t e^{-t^2}dt$	x or t	$\frac{2}{\sqrt{\pi}}\int_0^x e^{-x^2}dx$	$\frac{2}{\sqrt{\pi}}\int_0^t e^{-t^2}dt$
0.0	0.000000	0.0000	3.0	0.999977	0.9570
0.1	.112464	.0538	3.1	.999988	.9635
0.2	.222702	.1073	3.2	.999994	.9691
0.3	.328626	.1603	3.3	.999997	.9740
0.4	.428392	.2127	3.4	.999998	.9782
0.5	.520500	.2641	3.5	.999999	.9818
0.6	.603856	.3143	3.6		.9848
0.7	.677802	.3632	3.7		.9874
0.8	.742102	.4105	3.8		.9896
0.9	.796908	.4562	3.9		.9915
1.0	.842700	.5000	4.0		.9930
1.1	.880206	.5419	4.1		.9943
1.2	.910314	.5817	4.2		.9954
1.3	.934008	.6194	4.3		.9963
1.4	.952286	.6550	4.4		.9970
1.5	.966106	.6883	4.5		.9976
1.6	.976348	.7195	4.6		.9981
1.7	.983790	.7485	4.7		.9985
1.8	.989090	.7753	4.8		.9988
1.9	.992790	.8000	4.9		.9990
2.0	.995322	.8227	5.0		.9993
2.1	.997020	.8433			
2.2	.998136	.8622			
2.3	.998856	.8792			
2.4	.999310	.8945			
2.5	.999592	.9082			
2.6	.999764	.9205			
2.7	.999866	.9314			
2.8	.999924	.9410			
2.9	.999958	.9495			

TABLE II.

If $\Sigma(e^2)^{\frac{1}{2}}$ represent the square root of the sum of the squares of the residuals of n observations, the probable error of an observation is

$$0.6745\Sigma(e^2)^{\frac{1}{2}}/\sqrt{n-1} = A\Sigma(e^2)^{\frac{1}{2}},$$

and the probable error of the mean of n observations is

$$0.6745\Sigma(e^2)^{\frac{1}{2}}/\sqrt{n(n-1)} = B\Sigma(e^2)^{\frac{1}{2}}.$$

n.	A.	B.	n.	A.	B.	n.	A.	B.
1	.0000	.0000	11	.2133	.0643	21	.1508	.0329
2	.6745	.4769	12	.2034	.0587	22	.1472	.0314
3	.4769	.2754	13	.1947	.0540	23	.1438	.0300
4	.3894	.1947	14	.1871	.0500	24	.1406	.0287
5	.3372	.1508	15	.1803	.0465	25	.1377	.0275
6	.3016	.1231	16	.1742	.0435	26	.1349	.0265
7	.2754	.1041	17	.1686	.0409	27	.1323	.0255
8	.2549	.0901	18	.1636	.0386	28	.1298	.0245
9	.2385	.0795	19	.1590	.0365	29	.1275	.0237
10	.2248	.0711	20	.1547	.0346	30	.1252	.0229

TABLE III.

USEFUL CONSTANTS.

	Numeric.	Reciprocal.	Logarithm.
Naperian base e.	2.71828	0.367879	0.43429
Modulus M.	0.43429	2.30259	$\bar{1}$.63778
$\sqrt{2}$	1.41421	0.70711	0.15051
π	3.14159	0.31831	0.49715
2π	6.28319	0.15915	0.79818
$\pi/2$	1.57079	0.63662	0.19612
$\sqrt{\pi}$	1.77245	0.56419	0.24857

INDEX.

[*The figures refer to the pages.*]

Amplitude, 119.
Authority, 8.
Average, 41.

Belief, 12.
Bessel's interpolation, 52.
Borda's Series, 63.

Causation, 26.
Cause and effect, 22.
Chauvenet's criterion, 87.
Computation, 17.
Conditioned equations, 90.

Deduction, 11.
Deductive formulæ, 94.
Differences, 46.
Differences of logarithms, 62.
Dimensions, 38.

Empirical formulæ, 104.
Energy, 19.
Equations of condition, 113.
Errors, 66.

Errors, law of, 74.
,,   of mean square, 77.
Experiment, 31.

Fallacies, 13.

General formulæ, 92.
General mean, 85.
Graphical methods, 97.

Heredity, 4.
Hopkinson's method, 106.
Hypothesis, 20.

Ideas, 1.
Individuality, 1.
Induction, 10.
Integral $\int e^{-x^2} dx$, 81, 122.
Interpolation, 51.

Lagrange's interpolation, 54.
Law of error, 74.
Laws of Nature, 19.
Least squares, 88.

Matter, 19.
Means, 71.
Mean, error, 76.
Median, 44.
Mensuration, 56.
Mode, 43.
Modulus, 75.

Newton's interpolation, 52.

Object, 1.
Observation, 31.
Ofisometer, 57.

Peirce's criterion, 86.
Phase, 119.
Planimeter, 57.
Precision, measure of, 75.
Premiss, 7.
Probable error, 77.

Quartile, 44.

Reasoning, 7.
Recording instruments, 101.
Rejection of observations, 84.
Residuals, 73.

Sensation, 3.
Sine-curve, 118.
Subject, 1.
Symbols, 12.

Tables, 122.
   ,,   use of, 60.
Theory, 20.
   ,,   of exchanges, 29.

Units, 36.

Wave-lengths, 55.
Weddle's rule, 58.
Weight of observations, 84.
Will, 22.

MACMILLAN AND CO.'S LIST.

BY THE SAME AUTHOR.

Elementary Chemical Arithmetic. With 1100 problems. By SYDNEY LUPTON, M.A. With a table. Second Edition. Globe 8vo. 4s. 6d.

Numerical Tables and Constants in Elementary Science. With a map and table. Fcap. 8vo. 2s. 6d.

Lessons in Elementary Physics. By BALFOUR STEWART, M.A., LL.D., F.R.S. With a coloured spectrum. New and enlarged Edition. Fcap. 8vo. 4s. 6d. QUESTIONS. By T. H. CORE. Pott 8vo. 2s.

Physics. By BALFOUR STEWART. With illustrations. Pott 8vo. 1s. [*Science Primers.*

An Exercise Book of Elementary Practical Physics for Organised Science Schools under the Department of Science and Art, Evening Continuation Schools, and Elementary Day Schools. Arranged according to the Head Masters' Association's Syllabus of Practical Physics. By RICHARD A. GREGORY, F.R.A.S. Fcap. 4to. 2s. 6d.

Physics Note Book. Descriptions and Laws in Letterpress, with Space for the Students' Drawings of Experiments. Fcap. 4to. 2s. 6d. net.

Examples in Physics. Containing over 1000 problems with answers and numerous solved examples. Suitable for candidates preparing for the Intermediate, Science, Preliminary, Scientific and other Examinations of the University of London. By D. E. JONES, B.Sc. (Lond.). Third Edition. Revised and enlarged. Fcap. 8vo. 3s. 6d.

Intermediate Course of Practical Physics. By Prof. A. SCHUSTER, F.R.S., and C. H. LEES, D.Sc. Cr. 8vo. 5s.

Questions and Examples on Elementary Experimental Physics, Sound, Light, Heat, Electricity and Magnetism. By BENJAMIN LOEWY, F.R.A.S., etc. Pott 8vo. 2s.

A Graduated Course of Natural Science, Experimental and Theoretical, for Schools and Colleges. By B. LOEWY. Part I. First Year's Course. Globe 8vo. 2s. Part II. Second and Third Year's Course. With sixty diagrams. Globe 8vo. 2s. 6d.

Practical Lessons in Physical Measurement. By ALFRED EARL, M.A. Illustrated. Cr. 8vo. 5s.

A Laboratory Course in Experimental Physics. By W. J. LOUDON, B.A., and J. C. M'LENNAN, B.A. 8s. 6d. net.

MACMILLAN AND CO., LIMITED, LONDON.

MACMILLAN AND CO.'S LIST.

An Introduction to the Study of Chemistry. By W. H. PERKIN, Jun., Ph.D., F.R.S., Professor of Chemistry in Owens College, and B. LEAN, D.Sc., Assistant-Lecturer and Demonstrator, Owens College. Globe 8vo. 2s. 6d.

Experimental Proofs of Chemical Theory for Beginners. By WILLIAM RAMSAY, F.R.S. New Edition. Pott 8vo. 2s. 6d.

WORKS BY IRA REMSEN,
Professor of Chemistry, Johns Hopkins University.

The Elements of Chemistry. For Beginners. Fcap. 8vo. 2s. 6d.

An Introduction to the Study of Chemistry (Inorganic Chemistry). Cr. 8vo. 6s. 6d.

Compounds of Carbon: an Introduction to the Study of Organic Chemistry. Cr. 8vo. 6s. 6d.

A Text-Book of Inorganic Chemistry. 8vo. 16s.

WORKS BY SIR HENRY E. ROSCOE, F.R.S.,
formerly Professor of Chemistry, Owens College, Manchester.

A Primer of Chemistry. Illustrated. With Questions. Pott 8vo. 1s.

Inorganic Chemistry for Beginners. Assisted by J. LUNT, B.Sc. Globe 8vo. 2s. 6d.

Lessons in Elementary Chemistry, Inorganic and Organic. With Illustrations and Chromolitho of the Solar Spectrum, and of the Alkalies and Alkaline Earths. New Edition, 1892. Fcap. 8vo. 4s. 6d.

Lessons in Organic Chemistry. By G. S. TURPIN, M.A., D.Sc. Globe 8vo. Part I. Elementary. 2s. 6d. [*Part II. in prep.*

Practical Inorganic Chemistry. By G. S. TURPIN. Globe 8vo. 2s. 6d.

Chemical Theory for Beginners. By LEONARD DOBBIN, Ph.D., Assistant in the Chemistry Department, University of Edinburgh, and JAMES WALKER, Ph.D., D.Sc., Professor of Chemistry in University College, Dundee. Fcap. 8vo. 2s. 6d.

A Junior Course of Practical Chemistry. By FRANCIS JONES, F.R.S.E., F.C.S., Chemical Master in the Grammar School, Manchester. With a Preface by Sir HENRY E. ROSCOE, F.R.S. With illustrations. Gl. 8vo. 2s. 6d.

Questions on Chemistry. A series of problems and exercises in Inorganic and Organic Chemistry. By F. JONES. Pott 8vo. 3s.

A Course of Quantitative Analysis for Students. By W. N. HARTLEY, F.R.S., Professor of Chemistry and of Applied Chemistry, Science and Art Department, Royal College of Science, Dublin. Globe 8vo. 5s.

A Series of Chemical Problems. With Key. By T. E. THORPE, F.R.S. New Edition. Fcap. 8vo. 2s.

MACMILLAN AND CO., LIMITED, LONDON.

www.ingramcontent.com/pod-product-compliance
Lightning Source LLC
Chambersburg PA
CBHW030401170426
43202CB00010B/1445